ISBN 978-1-330-20224-1
PIBN 10051557

AN ELEMENTARY MANUAL
OF
RADIOTELEGRAPHY AND
RADIOTELEPHONY

AN ELEMENTARY MANUAL

OF

RADIOTELEGRAPHY AND RADIOTELEPHONY .

FOR

STUDENTS AND OPERATORS

BY

J. A. FLEMING, M.A., D Sc., F.R.S.

UNIVERSITY PROFESSOR OF ELECTRICAL ENGINEERING IN THE UNIVERSITY OF LONDON
PROFESSOR OF ELECTRICAL ENGINEERING IN UNIVERSITY COLLEGE, LONDON
MEMBER AND PAST VICE-PRESIDENT OF THE INSTITUTION OF
ELECTRICAL ENGINEERS OF LONDON
FELLOW AND PAST VICE-PRESIDENT OF THE PHYSICAL SOCIETY OF LONDON
MEMBER OF THE ROYAL INSTITUTION OF GREAT BRITAIN
ETC., ETC.

WITH ILLUSTRATIONS

THIRD EDITION

LONGMANS, GREEN AND CO.

39 PATERNOSTER ROW, LONDON

FOURTH AVENUE & 30TH STREET, NEW YORK

BOMBAY, CALCUTTA, AND MADRAS

1916

PREFACE TO THE THIRD EDITION

The Author published in 1906, through Messrs. Longmans, Green and Co., a volume on the Principles of Electric Wave Telegraphy and Telephony, in which an attempt was made to provide a fairly complete treatment of the subject, not limited to mere descriptions of various so-called systems of wireless telegraphy, but explanatory of the scientific principles underlying Radiotelegraphy in general. It was, however, represented to him that a smaller manual on the subject, suitable for the use of students, practical operators, and the general reader, on the same lines, but somewhat more elementary, might meet with acceptance.

The Author, therefore, endeavoured to put together in the present volume the information most likely to be of use for this purpose. Where it has been deemed advisable to introduce some little mathematical reasoning to supplement the verbal descriptions, it is limited to the use of simple operations and expressions. For the proof of many of the formulæ given in this Manual, which require rather more extended mathematical discussion than can be given here, the reader must be referred to the Author's larger book above mentioned.

It is assumed, however, that any user of this Manual has a general acquaintance with the elementary facts of electrical science.

It has not been considered necessary to encumber the pages with many references to original papers or patent specifications, since the student who masters this elementary treatise will be able at once to take advantage of the more complete information and references given in advanced text-books.

The subject has now acquired a position of such importance in connection with naval and military signalling and marine inter-communication generally, that means are required for adequately teaching the subject to electro-technical students and to practical operators, and thus equipping them with the initial scientific information necessary to enable them to follow intelligently its practical development, and also fit them for extending their knowledge of it by the study of advanced books and original papers.

Although there is no want of books upon the subject, many of them are occupied to a large extent with historical matter, and expositions of electrical phenomena which are either unnecessary for the practical radiotelegraphist, or can be obtained from other text-books. Hence reference to these matters is as far as possible omitted in this Manual, and the information given is confined to that which is necessary to enable a student familiar with the elementary facts of electricity and magnetism to proceed to the study of more advanced treatises on the subject of Radiotelegraphy.

In conclusion the writer desires to express his thanks for permission to make use of diagrams and illustrations of apparatus to the following firms and gentlemen :—To Senatore G. Marconi, and Marconi's Wireless Telegraph Company for the loan of blocks illustrating Marconi apparatus and stations; to the Amalgamated Radiotelegraphic Company and the Kilowatt Publishing Company for blocks of the Poulsen apparatus; to the Electrician Publishing Company for the use of many illustrations which have appeared in *The Electrician* of late years; to Mr. W. Duddell, F.R.S., for some curves employed in Chapter IX.;

to the Editors of *The Electrical Review*, and to other firms for illustrations of radiotelegraphic apparatus.

In preparing for the press this third edition the Author has added such new matter as is necessary to bring the information up to date and to correct misprints or errors, but the general purpose of the book has not been otherwise altered or its scope as an elementary manual enlarged. It is hoped, therefore, that its utility as an introduction to larger treatises will still be maintained.

J. A. F.

THE PENDER ELECTRICAL LABORATORY,
UNIVERSITY COLLEGE, LONDON,
September, 1915.

TABLE OF CONTENTS

CHAPTER I

ELECTRIC OSCILLATIONS AND ELECTRIC RESONANCE

CHAPTER II

DAMPED ELECTRIC OSCILLATIONS

CHAPTER III

UNDAMPED ELECTRIC OSCILLATIONS

CHAPTER IV

ELECTROMAGNETIC WAVES

CHAPTER V

RADIATING AND RECEIVING CIRCUITS

CHAPTER VI

OSCILLATION DETECTORS

CHAPTER VII

RADIOTELEGRAPHIC STATIONS

CHAPTER VIII

RADIOTELEGRAPHIC MEASUREMENTS

CHAPTER IX

RADIOTELEPHONY

AN ELEMENTARY MANUAL OF
RADIOTELEGRAPHY
AND
RADIOTELEPHONY

CHAPTER I

ELECTRIC OSCILLATIONS AND ELECTRIC RESONANCE

1. High Frequency Alternating Currents and Electric Oscillations.
—Since the art and practice of radiotelegraphy and radiotelephony
involve the employment of electric currents which alternate or
change direction in their circuits very rapidly, it is necessary to
commence the study of the subject by considering some of the
general properties of alternating currents.

By means of various appliances, such as a dynamo, voltaic cell,
thermopile, or other source of so-called electromotive force, we
can produce in certain bodies, known as electrical conductors, a
state in which they are said to be traversed by an *electric current.*
We recognise the presence of a current
by the production of *heat* in the conductor and a *magnetic field* around it.
Exploring the space near a conductor
carrying an electric current, by means
of a freely suspended magnetic needle,
we find the latter sets itself so as to
indicate that round the wire there is a
distribution of magnetic flux along
closed lines embracing the wire. Experimentally this is best illustrated by passing a stout copper wire through a hole
in a card just large enough to let it
pass, and connecting the ends of the
wire to a powerful battery or dynamo
(see Fig. 1). If we then place a small
pocket compass on the card and move
it about, we shall find that the needle
places itself at every point transversely to the wire. If iron filings

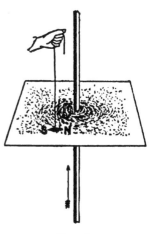

Fig. 1.

B

are sprinkled on the card and the latter gently tapped, the filings will be found to collect themselves more or less along certain circular lines, thus revealing the form and distribution of the invisible closed *lines of magnetic flux* round the conductor, whilst if the compass is placed on the card over the filings, it will be seen that the needle sets itself at all places so as to be in the direction of a tangent to these circles. If the connections of the wire with the terminals of the battery or dynamo are interchanged, it will be found that the compass needle reverses its direction, but the lines of magnetic flux remain circles as before. We may therefore speak of the current as having direction in the wire, since the magnetic field as indicated by the setting of the magnetic needle has direction one way or the other with reference to the wire. We are accustomed to call the direction of a line of magnetic flux the direction in which the north-seeking end, or, as it is usually called, the north pole N, of the compass needle points when placed on that line. The direction of the magnetic field is conventionally related to that of the current in the same manner as the twist and thrust of a corkscrew. Hence, if we imagine a watch laid face upwards on the above-mentioned card and that the circular lines of magnetic flux have the same direction as the rotating hands of the watch looked at from above, then the current creating them would be said to have a downward direction or to be flowing from the face to the back of the watch. If the direction of the magnetic flux remains constant from instant to instant, it is said to be due to a continuous, unvarying, or *direct current* (D.C.). If, however, the field changes its direction at regular intervals, being first right handed or clockwise in direction and then left handed or counter-clockwise, we say that the current is due to an *alternating current* (A.C.). The interval of time between two consecutive reversals of direction is called a *semi-period*, and the interval between two consecutive reversals in the same direction is called the *periodic time*, or *complete period*. The number of complete periods executed in a second is called the *frequency* of the alternations, and is generally denoted by the sign ∿. Thus 100 ∿ means that the frequency is 100, or there are 100 complete periods per second or 200 reversals of direction of the field and current per second.

If the frequency has any such value as 50 or 100 ∿ the current would be referred to as a *low frequency alternating current*. If, however, the frequency were of the order of 1000, 10,000, or 100,000, it would be described as a *high frequency alternating current*. There is, of course, no hard-and-fast line of demarkation. The terms high and low in this connection are relative or conventional.

When the frequency rises to a value of a million or so the current is generally called an *electric oscillation*.

In radiotelegraphy we are chiefly concerned with high frequency currents or electric oscillations of a frequency between 100,000 or so and 1 or 2 million, and we have first to consider the mode of representing them and their peculiar qualities.

It is most convenient to delineate an alternating current by means of a wave diagram, as follows :—

Draw a straight horizontal line and let distances marked off from one end represent *time.* Set up perpendicular (dotted) lines either above or below the line at equidistant points, the length of these lines representing the current in the circuit at that instant, the lines being drawn upwards when the current is in one direction and downwards when it is in the other direction. The gradual increase and decrease of the alternating current first in one direction and then in the opposite direction is then represented by the ordinates of an undulating curve, as in Fig. 2. A curve of this

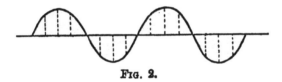

FIG. 2.

kind, called a sine curve, may be constructed in the following manner.

Take any line OX on which to mark off time (see Fig. 3), and with O as centre describe a circle with diameter AB. Divide its

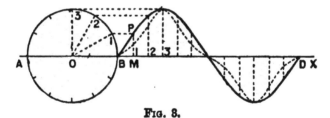

FIG. 3.

circumference into 12 parts, and through these points draw horizontal lines. Take any length BD on the horizontal line and divide it also into 12 parts and through these points draw lines perpendicular to OX to intersect the horizontal lines drawn through the 12 points on the circumference of the circle. Then mark dots at the intersections of the corresponding vertical and horizontal line, that is, at the intersection of horizontal line through point 1 on the circle with the vertical line through point 1 on the time line, and so on. Through the twelve intersection points draw a wavy curve. This curve is called a sine curve, because its ordinate PM is proportional to the sine of its abscissa MB, reckoning the whole length BD as divided into 360 parts or degrees. Thus if BM is $\frac{1}{12}$ of BD, the ordinate PM at that point M is proportional to the sine of the angle of 30°, or is 0·5 and on the same scale that the maximum ordinate of the curve is taken as unity. This maximum value of the ordinate is called the *amplitude* of the wave curve. An alternating current which is represented by such a sine curve is called a simple harmonic or simple periodic current.

We may, however, have alternating currents represented by any form of wave curve provided it is single valued, that is, has only one value of the ordinate for one given abscissa; in other words, by any periodic curve which does not cut or double back on itself.* In connection with alternating currents we are sometimes concerned with the maximum value or amplitude at given moments, but more frequently with a mean value of a particular kind called the *effective* or, usually, the *root-mean-square* (R.M.S.) value, defined as follows :—

The rate at which a current is producing heat in a conductor is at any instant proportional to the square of the current, and also simply proportional to the true effective resistance which the conductor offers to that current. If, then, the conductor has such a form that its true resistance for continuous currents is the same as for the alternating current in question, the mean value of the heat produced in any time is proportional to the mean or average value of the square of the current during that time. Hence we may ask the following question. If an alternating current of any given wave form exists in a circuit, find the value of the continuous or unvarying current which will produce heat at the same rate in the same conductor. Suppose we have the wave form of the current given. Then if we draw many equidistant current ordinates during one complete period and square them, that is, multiply the number denoting their length by itself, we can set off a new curve whose ordinates drawn at the same instant represent the square of the instantaneous values of the varying current. This curve is represented in Fig. 3 by the dotted curved line. If, then, we take the mean of the value of these squares and the square root of this mean, we have the value of the continuous current which would produce heat in the conductor at the same rate. The square root of the mean of the squares of the various instantaneous values of the current at equidistant intervals of

* Various curves representing alternating currents of complex wave form are shown in Fig. 2, chap. ix.

time is called the root-mean-square or R.M.S. value of the alternating current. When, therefore, we speak of an alternating current of 1 ampere we mean a periodic current which would produce heat at the same rate as an unvarying or continuous current of 1 ampere, assuming that the effective resistance of the conductor is the same in the two cases. The importance of this proviso will be seen later on.

2. **Undamped and Damped Electric Oscillations.**—When an alternating current of very high frequency exists in a circuit and continues uninterruptedly, it is usually called a persistent or *undamped electric oscillation*. It may be represented by a regularly repeated curve (see lowest curve, Fig. 4). We are, however, concerned in radiotelegraphy with a kind of alternating current of very

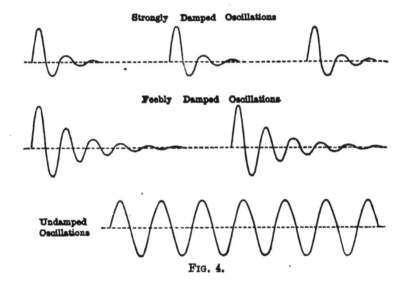

Strongly Damped Oscillations

Feebly Damped Oscillations

Undamped Oscillations

FIG. 4.

high frequency which consists of separate groups of alternating electric currents, each group beginning with the same amplitude, but then tailing away or damping down more or less quickly to zero, and after an interval of rest beginning again. Such a current may be represented graphically by the upper and middle curves in Fig. 4, and is spoken of as a series of trains of *damped electric oscillations*. In this last case we have therefore to consider:—

    (i.) The initial amplitude of each train.
    (ii.) The number of oscillations in each train.
    (iii.) The number of trains per second.
    (iv.) The rate at which the amplitude dies away in each train or the *damping*, as it is called.

As these damped electric oscillations are much used in radiotelegraphy, we shall consider more carefully some of their properties and the manner in which the above qualities are related to each other.

Suppose a pendulum to have a hollow bob with a hole at the bottom, the said bob being filled with ink or with sand. If the pendulum is set in vibration over a sheet of paper which is moved with uniform speed transversely to the direction of the plane of oscillation of the pendulum, the outflowing ink or sand would describe upon the paper a wavy line, gradually decreasing in amplitude as the vibrations of the pendulum die away (see Fig. 5).

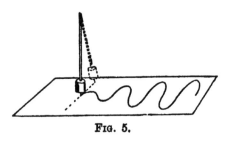

FIG. 5.

The vibrations of the pendulum each take place in the same time—that is, they are isochronous—and the interval of time between two movements of the bob in the same direction across its lowest position is called the periodic time of the pendulum. If the pendulum is slightly displaced from its position of rest, the bob is raised, and the action of gravity creates a *torque* or restoring couple, tending to bring it back again to its lowest position. If this torque and the corresponding angular displacement are both measured in suitable units, the quotient of the torque by the angle of displacement is called the torque per unit angle.

Again, if we suppose the whole mass of the pendulum divided into small portions, and the mass of each portion multiplied by the square of its distance from the axis of rotation, the sum of all such products for the whole pendulum is called its *moment of inertia*. It is shown in books on mechanics that the complete periodic time of a small vibration of a pendulum is obtained by dividing the square root of its moment of inertia by the square root of the torque per unit angle, and multiplying the quotient by the circular constant 2π or by 6·283. Accordingly, we may say that the time of vibration is given by the rule,

The complete periodic time of the pendulum

$$= 2\pi\sqrt{\frac{\text{Moment of inertia of pendulum}}{\text{Restoring torque per unit angle of displacement}}}$$

If the amplitude of the vibrations does not exceed a few degrees, then the time of vibration is independent of the amplitude. If then we were to measure the amplitude of successive vibrations as they die away, we should find that each amplitude bears the same relation to the preceding one in magnitude. Thus, suppose the first or initial displacement or amplitude is represented by the number 100, and the second one by 90, then the third one would be $\frac{9}{10}$ of 90, or 81, and the fourth $\frac{9}{10}$ of 81, or 72·9, and so on. Thus the numbers representing the successive amplitudes or excursions of the bob would be 100, 90, 81, 72·9, 65·6, 59·04, 53·14, 47·82, 43·04, 38·74, 34·86, etc.

These numbers are said to be in *geometrical* progression because each bears a constant ratio to the one preceding or following it in the series. If the logarithms of these numbers are written down, we obtain another series of numbers in *arithmetic* progression, successive terms having a constant difference. The reader is probably aware that there are two systems of logarithms in use, one to the base 10, which is employed in the construction of the ordinary slide rule, and the other called the Napierian system, to the base 2·71828. The Napierian logarithms are obtained from those to the base 10 by multiplying the latter by the modulus or factor 2·30259.

Accordingly, if we take the Napierian logarithms of the above series of numbers representing the gradually decreasing amplitudes of the pendulum as the vibrations die away, we have the following table:—

Amplitude of successive excursions which exhibit a constant ratio of 10 : 9.	Napierian logarithms of successive excursions which exhibit a constant difference 0·1053.
100·0	4·6052
90 0	4·4998
81·0	4·3945
72·9	4·2891
65·6	4·1838
59·04	4·0788
53·14	3·9730
47·82	3·8677
43·04	3·7622
38·74	3·6568
34·86	3·5514
etc.	etc.

The constant difference between the Napierian logarithms of the amplitudes of two successive swings is called the *logarithmic decrement* (abbreviated into *log. dec.*) of the motion. It is denoted by the symbol δ.

The reader should notice that we may speak of the log. dec. per complete period or the log. dec. per half period, according as the amplitudes plotted are the successive excursions of the pendulum in the same direction or in opposite directions. Thus in the above example the numbers 100, 90, etc., are the amplitudes of the successive swings in the same direction or to the same side. Hence the number 0·1053 is the log. dec. per complete period.

If the oscillations of the pendulum are permitted to die away, strictly speaking, this should take an infinite time, because each succeeding oscillation has a definite ratio to that of the preceding one. As a matter of fact, however, there will come a time when friction only just permits the pendulum to return to its zero position from its last small excursion, and the oscillations will then have ceased. We may, however, fix a definite useful limit to the final amplitude, and say that when the last excursion is so much reduced that it is only 1 per cent. or $\frac{1}{100}$ of the initial amplitude, the oscillations have for all practical purposes ceased. We can then easily find how many oscillations will take place before this happens. It is obvious that between the 1st and 10th swing in the same direction there are 9 complete oscillations, and between the 1st and 100th swing there are 99 complete oscillations. Again, in any system of logarithms the logarithm of 1 is always zero. Hence if we divide the number 4·6052, which is the Napierian logarithm of 100 by the log. dec., the quotient will give us a number which is one less than the number of complete swings in which the amplitude has been reduced to 1 per cent. of the initial amplitude. Thus in the case of the pendulum above mentioned we have 4·6052 ÷ 0·1053 = 43·7. Accordingly, in 44 to 45 complete swings the pendulum would be practically at rest, since the amplitude of its excursions would then have become reduced to 1 per cent. of the initial amplitude. The measurement of the logarithmic decrement enables us therefore to count the number of oscillations composing a train and therefore to say how fast they die away. A train of very few oscillations, say 5 or 6, is called a *highly damped train*, and a train of very many oscillations, say 100 or more, is called a *feebly damped train* (see Fig. 4).

These facts with regard to mechanical vibrations have their analogues in electric oscillations. We can by special devices, explained in the next chapter, set the electricity in certain forms of circuit in motion by giving it a sudden impulse or release. It then oscillates to and fro in the circuit, and creates rapid alternating currents, which, however, continually decrease in their maximum value because their energy is being dissipated by various causes such as resistance. Hence the oscillations are gradually damped out. We shall also consider in another chapter the manner in which this damping and the logarithmic decrement of the oscillations can be measured.

3. **Electric Circuits and their Qualities. High Frequency Resistance.**—Before discussing the production of electric oscillations we must refer to some of the qualities of electric circuits which are important in connection with high frequency currents. One of these is the *effective resistance* of the circuit. Electric resistance may be defined as the quality of an electric circuit in virtue of which the energy of an electric current existing in the circuit is dissipated as heat in the conductor. It may therefore be measured by the energy dissipated per second per unit current— that is, by the power dissipated per unit current. The practical unit of current is *the ampere,* which is defined as the unvarying current which, when flowing through a neutral solution of nitrate of silver between silver electrodes, deposits on the negative electrode 0·001118 gram of silver per second. The practical unit of power or work done per second is *the watt.* The practical unit of resistance (called *the ohm*) is therefore the resistance of a conductor which dissipates 1 watt as heat when a current of 1 ampere flows through it. The resistance of a conductor depends, however, upon the mode in which the current is distributed over the cross-section of it. Imagine a rod of copper of uniform section of any shape, and suppose it built up by laying together fine copper wires of square section placed parallel and closely packed. When a current flows through the rod we may picture to ourselves the current as uniformly divided between the small constituent wires, or we may suppose that these are insulated from each other and that some of the components carry more and others less current than the average, so that the total current is not distributed uniformly over the cross-section, but is denser in some places than in others. We can then very easily prove the following statement to be true. *The resistance of a conductor for uniform distribution of the current over its cross-section is less than that for any·non-uniform distribution of the same current.*

Let the large square in Fig. 6 be the section of the rod, and the small squares into which it is divided be the component elements. Let us suppose the total current to be equally distributed over the cross-section as indicated by the uniform shading. Then the conductor has a certain resistance, and dissipates a certain energy per second per unit of current flowing through it. In the next place, let us suppose that current is removed from one of the little elements and added to that in another, so that, whilst the current in one component element or wire becomes zero, represented by the small white square, that in the other selected constituent, represented by the doubly shaded square is doubled. This is in effect making the current non-uniform over the total section, without altering the total current flowing. The heat produced per second in any conductor by Joule's law is proportional to the square of the current, so that if the current is doubled, the heat is quadrupled. Hence, if we consider the energy dissipated as heat in each element or filament of the whole number into which we have considered the conductor to be divided, we see that when the current is uniformly distributed it is uniformly or equally dissipated as heat, and if there are, say, 1000 elements or little component conductors, then the total heat produced is 1000 times that in one element. If, however, we assume that the current is taken away from one element and added to that in some other element, then, as far as regards these two elements, the heat produced per second in the first is now zero, since the current is zero, and in the second the heat is quadrupled, since the current is doubled. In all the other elements it remains at the original value. Accordingly, although the total current flowing through the whole conductor is the same as before, the total heat generated has been increased by rendering the distribution of current over the cross-section non-uniform. From which it follows at once that the production of heat per unit current flowing through the conductor is a minimum for uniform distribution of the current over the cross-section.

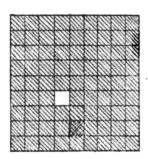

Fig. 6.

One of the particular characteristics of an alternating electric current, especially of a high frequency current, is that it is not distributed uniformly over the cross-section of the conductor, as is the case with direct currents, but is concentrated in a surface layer of the conductor or principally confined to the skin.

We may illustrate the difference between the two cases by a thermal analogy. Imagine an iron ball put into a furnace and left there for some time; it would become equally hot all through, and have the same temperature at the centre as at the surface. If, however, after being in the furnace for a short time, it is taken out and cooled, and then put into the furnace again, and these operations rapidly repeated, it would experience the changes of temperature only at the surface layers, and the interior would hardly change in temperature at all.

To make clear the reason for this superficial concentration of the current on conductors when it is rapidly alternating, we must consider a little more closely the manner and meaning of establishing a current in a conductor. For this purpose it is best to consider a simple case. Let there be two sheets of metal AB, CD (see Fig. 7), placed parallel to each other, and charged with electricities of opposite sign. These plates are shown in section in Fig. 7 and indicated by the thick black lines. This arrangement constitutes a *condenser*. The insulator or dielectric between the plates is in a peculiar state of strain along certain lines called *lines of electric strain*. In the diagram the directions of the strain at various parts is denoted by the dotted lines. The bodies we call conductors do not permit the creation in them of electric strain. If the strain in a dielectric exceeds a certain value, the insulator is ruptured and a spark discharge takes place.

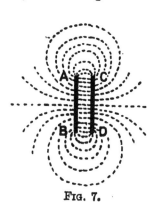

FIG. 7.

We may compare a dielectric with an elastic wire, which can endure a certain twist before breaking, whereas a conductor is like a thread of honey or some such plastic body, on which we cannot put any twist at all, because it yields immediately to the strain. The so-called charge of the condenser is the energy of this electric strain in the dielectric, and each cubic centimetre of the insulator stores up a certain amount of the whole energy.

For the sake of giving definiteness to our conceptions, we may think of the mass of the dielectric as divided up into closely compacted tubes, the sides of these tubes being bounded by lines in the direction of the electric strain. At the point where the tubes terminate on the conductors there is a positive or negative charge, and it is convenient to so select the size of these tubes that the charge on each end is a unit of electric quantity. We know that electricity can spread over the surface of conductors, but not over the surface of insulators. Accordingly, we must suppose that the ends of the tubes of electric strain are quite free to move over the surface of the conductors on which they abut.

In books on the theory of electricity, it is shown that the state of electric strain in a dielectric is equivalent to a tension or pull along the direction of the tubes of electric strain and a pressure at right angles to them. The whole mass of the dielectric may be considered to be in a state of stretch and squeeze, and in fact the attraction between oppositely electrified bodies is only the pull exerted by the tubes of electric strain extending between them.

Suppose, then, that the two oppositely electrified plates of the condenser in Fig. 7 are connected by a wire. The opposite ends of the tubes of electric strain move along it and approach each other, the tube shrinking up in the process in virtue of the tension along it. The disappearance of the tubes nearest the wire relieves the lateral pressure on others lying outside, and they all in turn collapse in the same manner. Each tube, however, represents a certain amount of potential energy stored in the dielectric, and it cannot disappear without leaving an equivalent behind it in some other form. The movement of positive and negative charges of electricity along the wire in opposite directions involves, therefore, the production of another effect in the dielectric, namely, a distribution of magnetic flux along closed curves embracing the wire constituting what we call the magnetic field due to electricity in motion. As, therefore, the energy of electric strain or the electrostatic field in the dielectric vanishes, due to the shrinking up of the tubes of strain, it is replaced by energy of magnetic flux or by a magnetic field distributed along endless lines enclosing the conductor.

This magnetic flux begins to be created at the surface of the conductor where the tubes of electric strain are vanishing, and it spreads outwards into the dielectric and also soaks or penetrates into the conductor much more slowly. It cannot yet be said that we understand fully the mechanism by which this energy transformation is effected, or how it is that the lateral movement of the tubes of electric strain which stretch from plate to plate, causes them to transfer their energy to another form in which it exists in a state called magnetic flux distributed along closed lines round the discharging wire.

In electromagnetic phenomena we recognise, however, that we are concerned with energy which may exist either in the form of electrostatic energy due to an electric strain in a dielectric produced by an electric charge, or with magnetic energy resulting from magnetic flux produced by an electric current.

The electrostatic form of energy has a close similarity to the potential energy of ordinary mechanical strain or distortion, and magnetic energy to the kinetic energy of moving masses. Electrostatic strain can exist in dielectrics, but not in conductors. Magnetic flux can exist both in dielectrics and in conductors, but the change of position of magnetic flux or movement of lines of magnetic flux through conductors causes a dissipation of some of the energy as heat.

Returning, then, to the case of the condenser which is being discharged, it must be understood that the lines of magnetic flux which replace the lines or tubes of electric strain do not spring into existence simultaneously at all parts of the field, but originate at the surface of the discharging wire and spread outwards into space, and also penetrate much more slowly into the wire itself, generating heat in the latter as they move through it.

If, then, the magnetic field outside the wire reaches its full and final state very quickly, the field inside the wire will have only penetrated by that time a very little distance into the metal. The speed with which the magnetic field outside the wire reaches its full development depends on the form of the circuit—that is, upon the *inductance* of the circuit.

The magnetic field, however, does not remain permanent, but in turn begins to disappear, its lines contracting in again upon the wire. It can be shown that this process recreates the electrostatic field, but with electric strain directed in the opposite direction to that strain, the collapse of which gave rise to the magnetic field. We shall consider this process more in detail in Chapter IV., when discussing electromagnetic waves. Meanwhile it will be sufficient if the reader is able to grasp clearly the following ideas. Let the small thick line circle in Fig. 8 represent the section of the wire discharging the condenser, and let the black dots round it represent the section of the lines or tubes of electric strain, and the fine line circles the direction of the lines of magnetic flux. Then, before discharge, we have in the dielectric only an electrostatic field, and the dielectric is permeated by lines or tubes of electric strain. When the discharge wire connects the plates of the condenser the electrostatic field begins to disappear, the lines of

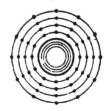

FIG. 8.

electric strain moving laterally in on the wire, but in so doing give up their energy to, and create closed lines of magnetic flux which move outwards from the wire. There comes an instant then when the electrostatic energy has all disappeared, the condenser is discharged, but a magnetic field has taken its place, and an electric current is said to flow in the discharge wire. This current then begins to die away—that is, the magnetic field collapses in upon the wire—but recreates as it disappears a fresh electrostatic field, the lines of magnetic flux fading away and being replaced by lines of electric strain in the opposite direction to the original strain.

If those transformations of energy succeed each other very rapidly in the space outside the discharge wire, then the magnetic field, which is more slowly soaking into the conductor, will never have time to penetrate far into it, and both the interior magnetic field, and therefore the heat evolution, will be confined to a thin outer layer of the material or to a mere skin. The better the conductivity of the material of which the discharge wire is made, the more slowly does the magnetic field travel into it, and therefore the thinner the skin.

These matters cannot be explained in full detail without the use of mathematical reasoning of an advanced character, but the above brief statement is perhaps sufficient to make it clear that when we are dealing with electric currents which change their direction very rapidly—in other words, with high frequency alternating electric currents—the current in the wire is not distributed uniformly over the cross-section of the wire, but is confined to a thin surface layer or skin. Hence, in accordance with what we have proved above, the resistance of the conductor for such alternating currents is not the same as that for steady or continuous currents, but is considerably greater. The usual tables for the resistance of copper and other wires give us what is called the ohmic or steady resistance, or resistance to continuous currents, of wires of various diameters and lengths. These values are not usually applicable in the case of conductors used in radiotelegraphy. We have, then, to consider the *high frequency resistance* of the wire. In the case of steady or direct currents, the resistance of a wire of given material, length, and cross-section varies directly as the length, inversely as the section, and directly as the *resistivity* or *specific resistance* of the material. These resistivities are given in books of reference either in absolute units or else in microhms or millionths of an ohm per centimetre cube. Thus, in the case of copper, the electric resistivity is 1·6 microhms per centimetre cube at ordinary temperature. If, then, we have a wire of 1 square millimetre in cross-section, or 0·01 square centimetre and 100 centimetres long, its steady resistance is $1·6 \times 100 \div 0·01 = 0·016$ ohm.

The ratio of the high frequency to the steady resistance has been investigated by many physicists. Lord Rayleigh has given two formulæ which enable us to calculate the resistance of a straight conductor for high frequency currents applicable in cases of moderate and very high frequency respectively, to straight or but slightly curved wires of any material. We have transformed his formulæ to meet the case of straight, round sectioned wires, as these are nearly always used in practice. Also, we have considered the wire to be of a non-magnetic material, say copper, as these are mostly employed in practice.

Let the letter c denote the circumference of the round wire measured in centimetres, n the frequency of the current, and ρ the steady current resistivity of the material of which the wire is made, reckoned in absolute units on the centimetre, gramme, second system of measurement. Thus for a copper wire $\rho = 1600$ at ordinary temperatures. We then calculate the value of the quantity $nc^2/\rho = k$. Thus if $n = 1000$, $c = 1$, and $\rho = 1600$, we have $k = \frac{5}{8}$. Let R' be the high frequency resistance of the wire and R its ordinary or steady resistance. Then, *provided that k is less than unity*, we can calculate the value of the ratio $\dfrac{R'}{R}$ from the formula

$$\frac{R'}{R} = 1 + \frac{k^2}{48} - \frac{k^4}{2880}$$

Thus for a copper wire of 1 centimetre in circumference or about $\frac{1}{8}$ inch diameter, and for a frequency of 1000 $\sim$, we have

$$\frac{R'}{R} = 1 + \tfrac{1}{123} - \tfrac{1}{18874} = 1\cdot008 \text{ nearly}$$

Accordingly, the increase in resistance for this wire and frequency is rather less than 1 per cent.

If the frequency and diameter of the wire are such as to make the value of the quantity k *very much greater than unity*, say at least equal to 6, then Lord Rayleigh has given another formula which, as modified by Dr. A. Russell, is equivalent to

$$\frac{R'}{R} = \tfrac{1}{2}\sqrt{k} + \tfrac{1}{4}$$

Thus, suppose the frequency $n = 1,000,000$ and the wire is a copper wire of one centimetre in circumference, then $\rho = 1600$, $c = 1$, and $n = 10^6$.

Hence $k = \tfrac{1000000}{16}$, and $\sqrt{k} = 25$, and $\tfrac{1}{2}\sqrt{k} = 12\cdot5$

Therefore the resistance of this wire for currents of 1,000,000 per second would be $12\frac{1}{2}$ times greater than its ordinary or steady resistance. Suppose the wire to be of one-tenth the diameter, or $0\cdot1$ centimetre in circumference, equivalent to about $\frac{1}{80}$ of an inch in diameter. Then for $n = 1,000,000$, $\rho = 1600$, and $c = 0\cdot1$, we have

$$k = \tfrac{100}{16} = 6\tfrac{1}{4}, \quad \text{and} \quad \sqrt{k} = 2\cdot5$$

Therefore $\dfrac{R'}{R} = 1\cdot25$, or the high frequency resistance of the wire is *approximately* 25 per cent. greater than its steady resistance. If, however, we consider the case of a copper wire still smaller, say of 0·01 centimetre in circumference, then for a frequency of 1,000,000 this would make $k = \frac{1}{16}$, and the last formula is then no longer applicable, since k is now less than unity, and we must revert to the previous formula. Accordingly for this last case we have

$$\frac{R'}{R} = 1 + \tfrac{1}{48}(\tfrac{1}{16})^2 - \tfrac{1}{2880}(\tfrac{1}{16})^4 = 1 + \tfrac{1}{12300}$$

Hence for so small a wire the resistance is not perceptibly increased even when using frequencies of a million and upwards.

The conclusion from the above statements is that, when using rather thick copper wires to convey high frequency currents, the effective resistance may be very much greater than it is for continuous or low frequency currents. On the other hand, if the wire is of very small diameter, then even for high frequency currents this increase in resistance due to the concentration of the current at the surface of the wire is not very serious. Accordingly, conductors for conveying high frequency currents should not be solid metal thick wires but stranded conductors, made by bunching or twisting together very numerous fine silk or copper . covered wires not thicker than No. 32 or No. 36 S.W.G. in size.

There is, however, another point in connection with this matter of equal importance. When a wire is coiled into a helix of many close turns it is found that its high frequency resistance is considerably greater than that of the same wire stretched out straight, even as calculated by Lord Rayleigh's formula. The reason for this is as follows :

We have already shown that anything which causes a non-uniformity of distribution of current over the cross-section of the wire causes an increase in resistance. In the case of a straight wire we have also shown that high frequency currents concentrate themselves near the surface of the wire. If the wire is straight, this surface layer of current is uniformly thick all round the section of the wire. If, however, the wire is coiled into a spiral, the current is not merely concentrated in a surface layer, but is furthermore concentrated on the inner side of each turn, so that the surface layer of current is of different thicknesses at different parts of the circumference (see Fig. 9). From this it follows that we have an increase in the resistance to high frequency currents as compared with the resistance of the same wire when stretched out straight. This effect has been investigated experimentally by Dolezalek, Battelli, and Magri, and by T. B. Black, and theoretically by Wien, Sommerfeld, and by L. Cohen. Battelli and Magri, as well as Black, placed equal lengths of two similar wires in the glass bulbs of two similar air thermometers, one of the wires being straight and the other tightly coiled. A high frequency current or train of oscillations was then sent through the two wires in series, and the amount of heat generated in the two cases measured by the increase of air pressure created in the bulbs. Using two sizes of wires about 0·15 and 0·3 cm. in diameter, and oscillation frequencies of 1,000,000 and 5,000,000 respectively, Black

FIG. 9.

found that the resistance of the wire in the form of a spiral was to that of the same wire stretched out straight in some ratio between 1·25 and 1·89. Hence we may say that the effective resistance of a wire for the above frequencies may be increased from 25 to 90 per cent. by coiling it in a helix. This is over and above the increase in resistance which results from the surface disposition of the current as compared with the resistance to steady currents. Wien and Sommerfeld have investigated the matter theoretically. Sommerfeld considered the case of a tightly wound spiral or solenoid of one layer made up of square-sectioned wire with extremely thin insulation, and he deduced from this an expression for the ratio of the high frequency resistance R'' of a spiral of one layer of circular-sectioned wire to the resistance to continuous currents R which is equivalent to

$$\frac{R''}{R} = 2\sqrt{\pi}\,\tfrac{1}{2}\sqrt{k}$$

where k has the same meaning as in the Rayleigh formula given above. The constant $2\sqrt{\pi} = 3\cdot54$ and $\tfrac{1}{2}\sqrt{k}$ is the value of the ratio of the high frequency resistance R' of the wire stretched out straight to the steady current resistance R. Black's experiments show, however, that the actual ratio is not much more than half the above value.

L. Cohen has also given a formula for the ratio of the resistance of a helix to that of the same wire when stretched out straight when the frequency is of the order of 10^6 and the diameter of the wire (d) not extremely small, say 1 mm., which is equivalent to

$$\frac{R''}{R'} = \frac{8N^2d^2}{\pi}\left(1 + \tfrac{1}{9} + \tfrac{1}{25} + \text{etc.}\right)$$

c

where N is the number of turns per unit length of the helix. This formula, however, appears to give ratios which are greater than those furnished by experiment.

Black found that for very high frequencies the ratio $\frac{R''}{R'}$, viz. that of the helix to that of the same wire stretched out straight, was independent of the frequency when the frequency was very high, but determined by the number of turns per unit of length and by the diameter of the wire, and also that it was also determined by the ratio of length to mean diameter of the helix, at least up to a ratio of 10, beyond which, however, all spirals behave like infinite spirals.

None of the formulæ so far given by theory appear, therefore, to be capable of predicting very accurately the high frequency resistance of such solenoids or coils as are used in radiotelegraphy.

It is, however, important to bear in mind that for wires of the diameter of 1, 2, or 3 mm., and for frequencies of 10^6 or so, the resistance of the wire when stretched out straight will be about four, eight, or twelve times that of the steady or ordinary resistance, and that when coiled into spirals of eight or ten turns per inch this augmented resistance may be again increased by 50 to 100 per cent., that it may become half as much again, or double that which it is for the same wire subjected to the same oscillations, but stretched out straight. It is, therefore, difficult to predict the power dissipated as heat by electric oscillations in spirals made of wire not very thin.

4. High Frequency Inductance.—The inductance of a circuit is that quality of it in virtue of which energy is conserved or associated with the circuit when a current flows through it. There is a very close analogy between the dynamical quantity called the mass of a body and the inductance of a circuit, and it may be that the so-called mass of matter is in fact due to the same ultimate qualities which create inductance.

If an ordinary body is in motion without rotation, with a certain linear velocity v, then it possesses a certain store of kinetic energy. This energy is measured by the product of half the square of the velocity and a certain constant, M, called the mass of the body, or by $\frac{1}{2}Mv^2$. In the same manner, if an electric current i is circulating in a circuit it involves a certain storage of energy which is proportional to half the square of the current and to a certain constant, L, called the *inductance* of this circuit, or to $\frac{1}{2}Li^2$.

Again, if a heavy body is in rotation round an axis it has a certain angular velocity, ω, at any instant, and a certain angular energy. This last is proportional to half the square of the angular velocity, and to a coefficient, K, called the *moment of inertia*, or to $\frac{1}{2}K\omega^2$.

Furthermore, if the linear or angular energy and velocity of a body are changing, the rate at which the energy changes with the velocity is called the linear or angular momentum (Mv or $K\omega$), and if the momentum is changing with time, the time rate of change of the linear or angular momentum is called the force or torque acting on the body. In the case of an electric circuit, if the energy stored up is changing with the current, then its rate of change (Li) is called the electrokinetic momentum of the circuit. If this latter is changing with the time, the time rate of change of the electrokinetic momentum is the electromotive force acting in the circuit due to inductance. Having regard to the fundamental discovery of Faraday that a change in the number of lines of force linked with or passing through a circuit gives rise to and is a measure of the electromotive induced in it, we easily see that the electrokinetic momentum Li is merely another name for the total magnetic flux due to the current in a circuit which is self-linked with the circuit.

The quantity called the inductance may, therefore, be considered to be a coefficient or number which denotes the number of lines of magnetic flux which are self-linked with a circuit when unit current flows through it. If the unit of current is the *ampere*, then the corresponding unit of inductance is called the *henry*, and a circuit having an inductance of 1 henry when traversed by a current of 1 ampere produces a total self-linked magnetic flux which is generally called 1 *weber*. If this flux is removed uniformly from the circuit in 1 second it would generate in it an electromotive force of 1 *volt* whilst the removal lasts.

The inductances of such circuits as we have to deal with in radiotelegraphy are generally small, and are conveniently measured in *millihenrys* or *microhenrys*, these being respectively one-thousandth and one-millionth of a henry.

The above statements may be perhaps made clearer by considering a concrete example.

Imagine a wire bent into a circle to have some source of electromotive force, say a battery, inserted in it of such magnitude as to make a current of 1 ampere circulate round it. The current will create a magnetic flux which is linked with its own circuit, and if the inductance of this circular wire is 1 henry, then a total flux of 1 weber will be self-linked with the circuit. A flux of 1 weber is equal to 100 million "lines of force" in the usual mode of reckoning. Let the battery be short circuited, then the current in the circuit commences to die away, and in so doing removes magnetic flux from the circuit. This action creates an induced electromotive force in the circuit acting in the same direction as the battery. If the flux is removed uniformly in 1 second, and if the constant induced electromotive force during that time is equal to 1 volt, then the total initial self-linked flux has a value called 1 weber, and the inductance of the circuit is 1 henry. The inductance of a circuit therefore depends on its geometrical form, and is determined by the amount of linkage of flux that takes place with itself when 1 ampere flows in the circuit. The inductance is therefore increased by closely coiling the circuit. Also it is determined by the mode in which the current is distributed over the cross-section of the conductor and is therefore not the same for steady currents uniformly distributed as for high frequency currents which are concentrated at the surface of the conductor. The actual calculation of the inductance of any conductor is impossible, or very difficult, except in a few particular cases in which the circuit is of a simple form, such as a straight wire with return at an infinite distance, a circular, square, rectangular, or spiral circuit. The principle on which such predetermination proceeds is that of calculating the work done in dividing up the current into a series of elementary filaments of current, and removing them all to an infinite distance from each other. We may consider the whole current to be divided into a number of parallel streamlets or filaments of current in the conductor. Now, parallel currents moving in the same direction attract each other. Hence, if we assume that these filaments of current can be separated from each other to an infinite distance, as if we were separating a rope into its constituent strands, the work so done in taking the current to pieces must be equal to the energy it possesses as a whole before separation. This last is equal to $\frac{1}{2}Li^2$, where i is the whole current and L the inductance of the circuit. The quantity L is therefore sometimes called the potential of the circuit on itself. It was shown by Neumann that the potential of a circuit on itself is obtained by summing up all the products obtained by multiplying together the lengths of all possible pairs of small elements of the circuit multiplied by the cosine of the angle between their direction, and divided by their distance apart, regard being taken in this process of the distribution of the current over the cross-section. The actual calculation is even in the simplest cases somewhat difficult, and for details the reader must be referred to larger treatises, but we shall here give the formulæ for some cases often required in practice. These expressions, with the exception of the formula (6) for the inductance of a helix of one layer, are the high frequency values applicable in cases in which the current is confined simply to a surface layer of the conductor.

In these formulæ we suppose the circuit to consist of a circular-sectioned wire of diameter d centimetres and of length l centimetres, made of a non-magnetic material, the diameter being small compared with the length.

The inductance L in these formulæ is expressed in microhenrys, and this can be reduced to centimetres or absolute electromagnetic units by multiplication by 1000.

(1) *Inductance of a straight wire of length l centimetres.*

$$L = \frac{2l}{1000}\left\{2\cdot3026 \log_{10}\frac{4l}{d} - 1\right\} \text{ microhenrys}$$

The above formula is applicable in calculating the inductance of a single wire antenna of copper or other non-magnetic wire when we are concerned with high frequency oscillations in it.

(2) *Inductance of a square circuit formed of circular-sectioned wire.*

Length of side of square $=$ S centimetres.

$$L = \frac{8S}{1000}\left\{2\cdot3026 \log_{10}\frac{16S}{d} - 2\cdot853\right\} \text{ microhenrys}$$

(3) *Inductance of a circular circuit formed of circular-sectioned wire.*

Mean diameter of the circle $=$ D centimetres.

$$L = \frac{2\pi D}{1000}\left\{2\cdot3026 \log_{10}\frac{4\pi D}{d} - 2\cdot45\right\} \text{ microhenrys}$$

The above formulæ (2) and (3) are applicable for calculating the high frequency inductance of a square and circular circuit respectively.

(4) *Inductance of two parallel wires D centimetres apart, length of each wire being l centimetres.*

$$L = \frac{4l}{1000}\left\{2\cdot3026 \log_{10}\frac{2D}{d}\right\} \text{ microhenrys}$$

(5) *Inductance of a rectangular circuit of circular-sectioned wire. The sides of the rectangle are respectively A and B centimetres, and the diagonal D centimetres.*

$$L = \frac{9{\cdot}2104}{1000} \left\{ (A + B) \log_{10} \frac{4AB}{d} - A \log_{10}(A + D) \right.$$

$$\left. - B \log_{10}(B + D) - \frac{A + B - D}{1{\cdot}1513} \right\} \text{ microhenrys}$$

(6) *Inductance of a helix of one single layer of circular-sectioned wire.*

N = number of turns per centimetre of length of the helix.
D = diameter of helix in centimetres.
l = length of helix in centimetres.

$$L = \frac{(\pi DN)^2}{1000} l \left\{ 1 - 0{\cdot}424\left(\frac{D}{l}\right) + 0{\cdot}125\left(\frac{D}{l}\right)^2 \right.$$

$$\left. - 0{\cdot}0156\left(\frac{D}{l}\right)^4 \right\} \text{ microhenrys}$$

In making actual concrete standards of inductance the formulæ (2), (5), and (6) will be found very useful.

A very fairly accurate standard of inductance may be made as follows :—Chuck in a lathe a rod of ebonite or hard dry wood and turn it truly cylindrical. Then with the screw-cutting gear cut a screw groove in it of 8 or 10 turns to the inch, and wind up in this groove bare copper wire of No. 14 or No. 16 S.W.G. size. The length of the helix may be from 10 to 50 times its diameter. The ends of the copper wire should be secured to terminals. Then measure the mean diameter of the helix of wire, taking into account the thickness of the wire. This gives the value of D in the above formula. The length of the helix (l) is then measured and the number of turns (N) per unit of length, and these quantities inserted in the formula (6) will give the inductance of the helix in microhenrys, each of which is 1000 absolute electromagnetic units. Since the ratio $\frac{D}{l}$ appears in the formula (6), we may measure the diameter and length of the spiral in inches if more convenient, and also count the turns per inch. It will then be seen that the quantity (πDN) which appears in the formula is the length of wire wound on one unit of length of the helix, and is the same number whether D is measured in inches and N in turns per inch, or D is measured in centimetres and N in turns per centimetre. Hence, if we adopt the inch as the unit of length, the only change we have to make is to insert for the value of l in the formula (6), where it appears alone in front of the bracket, the length of the helix measured in centimetres.

The above formula is due to Dr. A. Russell, and is of considerable practical value when using currents of low or moderate frequency, but when high frequency currents are employed, the actual inductance of the spiral will be slightly less than that given by the formula by an amount which can be predetermined theoretically by a formula due to L. Cohen.

It is necessary always to bear in mind that the inductance of a coil or wire depends to some extent upon the manner in which the current is distributed over its cross-section. The resistance of a wire, we have shown, is a minimum for uniform distribution of current over the cross-section and is increased by any cause tending to make it non-uniform. On the other hand, the inductance is a maximum for uniform distribution over the cross-section, and is diminished by any cause tending to make it non-uniform. Hence it is less for high frequency currents than for steady currents. Also, in the case of coiled wires, the increase in frequency has a greater effect in diminishing the inductance than is the case with straight wires. Dolezalek has suggested constructing standards of inductance of bunched or stranded insulated wires, each wire being not more than 0·1 mm. diameter, or, say, No. 40 S.W.G., to compel uniformity in the current distribution over the section.

5. **Electrical Capacity.**—If we communicate a charge of positive electricity to a conductor, we raise its potential, that is, we increase the quantity of work required to be done to transfer a unit of positive electricity carried on a small conductor from the earth to the surface of the conductor in question. The charge which must be given to a conductor to raise its potential by one unit is a measure of its *electrical capacity*. If this capacity is denoted by the letter C, and the potential of the conductor by V, then the quantity of electricity on it is Q, where $Q = CV$.

The practical unit of electric quantity is the *coulomb*, and the practical unit of potential is the *volt*, and the practical unit of capacity is the *farad*. Hence, to charge a capacity of 1 farad to a potential of 1 volt, we must place on it a charge of 1 coulomb of electricity. Faraday showed that it is impossible to create a charge of electricity of one kind only. Hence, if any conductor is charged with 1 coulomb of positive electricity, there must be an equal charge of negative electricity on some other conductor or conductors, or on the earth.

A *Condenser* is any arrangement of two conductors on one of which a charge of positive electricity can be placed, and on the other an equal quantity of negative. The magnitude of either quantity is called the charge of the condenser. The potential difference (P.D.) between the conductors is called the terminal

P.D. of the condenser. The quotient of charge by terminal P.D is called the capacity of the condenser. In radiotelegraphy the capacities with which we are concerned are generally very small and best measured in microfarads (*mfds*), that is, in millionths of a farad, or in micro-microfarads (*mmfds*), that is, in millionths of millionths, or billionths of a farad.

In the case of certain conductors of symmetrical form we can predetermine the capacity from the geometrical form and dimensions.

Thus, for instance, consider the case of a conducting sphere placed in free space at a great distance from all other conductors. If we place on this sphere a charge of electricity, it will distribute itself equally. It is a fundamental principle that all parts of a conductor must be at the same potential, and it can be shown from first principles that the potential in electrostatic units at any point at a distance x from a charge of positive electricity q, also reckoned in electrostatic units, collected on a very small sphere, is $\frac{q}{x}$.

Hence, in the case of the charged sphere the charge on its surface may be divided into small elements of charge which are all equidistant from its centre. Hence, if the radius of the sphere is R and its whole charge is Q, its potential is $\frac{Q}{R}$.

But its capacity C is equal to the quotient of charge by potential. Hence its capacity is numerically equal to its radius when reckoned in electrostatic units.

The electrostatic unit of potential is equal to 300 volts. The electrostatic unit of quantity is such that 3000 million electrostatic units are equal to 1 coulomb.

Accordingly 900,000 million electrostatic units of capacity must be equal to 1 farad, or 900,000 to 1 microfarad, or 0·9 of a unit to 1 micro-microfarad.

Accordingly, we have the following rule:—

To reduce capacities reckoned in electrostatic units to their equivalent reckoned in microfarads, divide by 900,000.

Hence if a sphere has a radius R centimetres, its electrostatic capacity is R electrostatic units and its capacity in microfarads is $\frac{R}{900000}$.

For convenience of reference we give below the formulæ for the capacity of conductors or condensers of various forms expressed in microfarads (mfds).

(1) *The capacity of a sphere of radius R centimetres when at a distance from all other conductors and from the earth.*

$$C = \frac{R}{900000} \text{ mfds.}$$

(2) *The capacity of a thin, flat circular disc of diameter D centimetres when at a distance from all other conductors and from the earth.*

$$C = \frac{D}{3 \cdot 1415 \times 900000} = \frac{D}{2827431} \text{ mfds.}$$

(3) *The capacity of a condenser formed of two flat plates each of area A square centimetres placed parallel to each other at a distance d centimetres, d being small compared with $\sqrt{A}$.*

$$C = \frac{A}{4\pi d \times 900000} = \frac{A}{11309724 \times d} \text{ mfds.}$$

(4) *The capacity of a condenser formed of two thin concentric tubes l centimetres long and of diameters D_1 and D_2 centimetres respectively.*

$$C = \frac{l}{4 \cdot 6052 \log_{10} \frac{D_1}{D_2} \times 900000} \text{ mfds.}$$

(5) *The capacity of a vertical wire of length l centimetres and diameter d centimetres at a considerable distance from the earth and from all other conductors.*

$$C = \frac{l}{4 \cdot 6052 \log_{10} \frac{2l}{d} \times 900000} \text{ mfds.}$$

(6) *A horizontal wire of length l centimetres and diameter d centimetres stretched parallel to the earth's surface at a distance h centimetres above it.*

$$C = \frac{l}{4 \cdot 6052 \log_{10} \frac{4h}{d} \times 900000} \text{ mfds.}$$

Concerning the above formulæ a few remarks are necessary. In the first place the assumption made in all cases is that the dielectric medium surrounding or between the conductors is air or some other material of unit dielectric constant. If, for instance, in the formulæ (3) and (4) the dielectric between the plates or cylinders is glass, ebonite, oil, or other material, the formula must have the constant K prefixed as a multiplier, K denoting the dielectric constant of the material as taken from the following table :—

DIELECTRIC CONSTANTS OF VARIOUS INSULATORS
FOR AIR K = 1.

Dielectric.	Dielectric constant K at 15° C.
Glass, flint	6·57 to 10·1
Glass, crown	6·96
Ebonite	2·05 to 3·15
Indiarubber, pure	2·12
Indiarubber, vulcanised	2·69
Mica	6·64
Sulphur	2·9 to 4·0
Shellac	2·7 to 3·0
Paraffin oil	2·00
Turpentine	2·23
Benzol	2·38

The values vary a good deal for different specimens of materials of the same name, especially in the case of solids of complex composition.

In the next place it should be noted that the formulæ for the plate condenser (3) and cylinder condenser (4) have been obtained on the supposition that the lines of electric strain stretching between the conductors are straight lines. As a matter of fact, at the edges they are curved and there is a correction for the fringe of lines of force at the edges which has been omitted. It is not large when the plates are near together.

In the third place the formulæ for the capacity of the sphere (1), the disc (2), and the vertical wire (5) have been obtained on the supposition that they are removed a long way from the earth. In the case of actual conductors of this kind, the measured capacity would be found to be greater than that given by the above formulæ. The manner in which this capacity can be experimentally obtained will be described in a later chapter.

In those cases in which a condenser has to be constructed of definite and measurable capacity for radiotelegraphic purposes the best type is a condenser consisting of flat metal plates (sheet tin) immersed in paraffin oil. If two such plates are placed in oil separated by a distance very small compared with their linear dimensions, the capacity can very approximately be calculated by the formula (3) and the dielectric constant may be taken as equal to 2·0, and a condenser of predetermined capacity thus made.

For many purposes, such as small-power transmitters, the familiar Leyden jar, consisting of a glass bottle or tube coated with tinfoil or silvered or coppered on a portion of its interior and exterior surfaces, is still much used as a condenser.

The student should note that any insulated conductor in the neighbourhood of the earth may be considered to be the inner conductor or coating of a form of condenser of which the earth is the other coating or conductor, and the air the dielectric. Hence an aerial wire as described in Chapter V. is such a condenser, and has capacity with respect to the earth, the value of which for a single isolated wire removed by a considerable distance from the earth has been given on a previous page.

It is important to notice the mode in which capacities can be added in under various circumstances. Suppose we have a number of plate condensers consisting of metallic plates close together, but separated by a dielectric, such, for instance, as a Leyden jar. Then when these plates have received electric charges, one positive and the other negative, lines of electric force proceed from one plate to the other, and except at the edges of the plates these lines are short and straight.

If then we join two or more such condensers in parallel by uniting by wires all the outside coatings and also all the inside coatings, the disposition of the lines of electric force passing from one plate to the other is not much affected by the proximity of the different condensers. Hence the joint capacity of all the condensers in parallel is the sum of each of them separately. On the other hand, if the separate condensers are joined in series or in cascade, the reciprocal of the capacity of all in series is the sum of the reciprocals of each separately.

It must be noted that these rules do not apply in cases in which the proximity of the separate condensers disturbs the previous distributions of their electric fields. Thus, for instance, we could not put two spheres in contact and say that the capacity of the two together is the sum of each of them separately, because the mere fact of putting the spheres together entirely changes the simple uniform distribution of electricity on each of them separately when placed far apart.

The same thing holds good for a number of aerial wires. A single vertical insulated aerial wire has a certain capacity, but if we place a number of such wires at all near together and parallel to each other, we cannot say that the capacity of the whole lot is the sum of each separately. It is, indeed, far less because the wires shield each other and disturb each other's free electric field. Professor Howe has given formulæ which enable us in such cases to calculate the joint capacity of a number of aerial wires stretched parallel to each other through the air.

Let there be a number of circular-sectioned wires the radius of the section or semi-diameter being r cms., and let each wire be l cms. long. Suppose n of these wires are stretched parallel in the air at a distance d cms. apart, the wires being at a considerable height above the earth, as in the case of a parallel wire antenna. Then Howe has given the following formula for the approximate joint capacity of the wires provided l is large compared with nd :—

$$C = \frac{nl}{2\left\{n\left(\log_e \frac{l}{d} - 0\cdot309\right) + \log_e \frac{d}{r} - B\right\}}$$

where B is a constant depending on n as follows :—

n	B	n	B
2	0	7	4·85
3	0·46	8	6·40
4	1·24	9	8·06
5	2·26	10	9·80
6	3·48	11	11·65
		12	13·58

The above formula gives the capacity C in electrostatic units, and can be reduced to microfarads by dividing by 9×10^5.

To find the capacity of a single circular-sectioned wire of length l and diameter $d = 2r$ we take as origin the centre of the wire. Let x be the distance of any other point on the axis and δx the thickness of a slice of the wire at that point. If ρ is the surface density then $2\pi\rho r \delta x$ is the surface charge of the slice, and hence $2\pi\rho r \delta x / \sqrt{r^2 + x^2}$ is its potential at the centre. Therefore, the whole potential V of the wire is given by

$$V = 4\pi\rho r \int_0^{l/2} \frac{\delta x}{\sqrt{x^2 + r^2}} = 4\pi\rho r \left\{\log_e\left(\frac{l}{2} + \sqrt{r^2 + \frac{l^2}{4}}\right) - \log_e r\right\}$$

But the whole charge on the wire $= Q = 2\pi\rho rl$ and $Q = CV$. Hence, if the length of the wire is large compared with its diameter its capacity C is given by

$$C = \frac{l}{2 \log_e \dfrac{l}{r}} \text{ (in electrostatic units).}$$

Using ordinary logarithms and microfarads this becomes as on page 25—

$$C = \frac{l}{4 \cdot 6052 \times 900000 \times \log_{10} 2l/d}$$

6. The Time-period of Oscillatory Electric Circuits.—If we join in series with each other some form of inductive circuit and some form of condenser, we have a circuit which is called an *oscillatory electric circuit*, and such circuits are broadly divided into two classes, respectively called *open* and *closed oscillatory circuits*.

Thus, suppose we bend a thin wire into the form of a square and connect the two ends to two metallic plates placed very near to each other (see Fig. 10), we have a circuit which is called a *closed oscillatory circuit*. It possesses inductance owing to the size and form of the square circuit and capacity in virtue of the proximity of the two plates which form a condenser.

If, however, we affix a metal sphere or disc to the end of a long vertical wire which has its lower end in close proximity to but not touching the earth (see Fig. 11), we have an *open oscillatory*

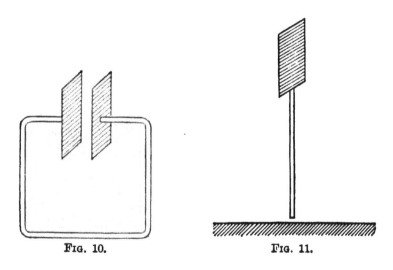

Fɪɢ. 10.   Fɪɢ. 11.

circuit which possesses inductance in the wire and capacity since the sphere or disc and the wire itself forms with the earth the two surfaces of a condenser. Between these extreme types of oscillatory circuit we can have many others, all characterised by possessing inductance and capacity in series. If the electric charge in such a circuit is disturbed in any way, as by introducing into the circuit a sudden electromotive force, this charge oscillates backwards and forwards with a definite and constant time period, just as a pendulum when disturbed and left to itself executes mechanical vibrations.

It will be necessary to consider a little more in detail how these oscillations are produced. Let us suppose the condenser has a capacity of C farads and that the circuit connecting its plates has an inductance L henrys and a resistance of R ohms. Then let the condenser plates be charged until they have a potential difference of V volts. Suppose that the circuit is now completed through the inductance, the condenser commences to discharge. Electricity passes in the form of a current round the circuit, and as the charge in the condenser diminishes the current in the circuit increases and the energy is transformed from electrostatic to electrokinetic form. At the outset the energy imparted to the condenser was equal to $\frac{1}{2}CV^2$ joules, and when the discharge is complete the result is that we have a current in the circuit say of A amperes, and therefore an electrokinetic energy $\frac{1}{2}LA^2$ joules associated with it, and also some portion of the original energy has been transformed into heat in consequence of the resistance of the circuit. This current energy, as already explained, expends itself again partly in reproducing electrostatic charge, and the transformation repeats itself again and again until the whole original energy is dissipated as heat in the circuit or by any other dissipative action. These electrical operations may be compared with the similar energy changes which take place in the case of mechanical vibrations. Imagine a heavy mass M hung up by a spiral spring. The spring can be stretched, and to stretch it work must be done on it and energy of strain stored up in it. If then the heavy bob is pulled down and released, the energy of strain stored up in the stretched spring expends itself in creating motion in the heavy mass. When the spring has returned to its original dimensions it has expended this energy in imparting a kinetic energy $\frac{1}{2}Mv^2$ to the bob, with the exception of that small part which may have been expended in making eddies in the air or in heating the metal of the spring. This kinetic energy, however, expends itself in turn in compressing the spring and then is again transformed into the form of energy of strain, and the process repeats itself again and again until the original store of energy is frittered away and the bob comes to rest.

Hence, just as the original energy represented by the stretch of the spring is transformed into energy of motion of the bob, and then back again, so the energy of electric strain originally created in the condenser expends itself in creating an electric current in the inductive circuit, and then this in turn re-creates the energy of electric strain, and the process repeats itself until the energy is frittered away into heat or removed by radiation. In the one case we have mechanical oscillations, consisting in the up-and-down motion of the heavy mass, and in the other electric oscillations consisting in a high frequency alternating current in the inductive circuit.

Suppose, then, that at any instant during the discharge of the condenser the current in the inductive circuit is a amperes, and at that moment the charge of the condenser is q coulombs, the maximum or original charge being Q coulombs. At that instant the potential difference of the condenser plates must then be $\frac{q}{C}$ volts. Again the fall in potential down the inductive circuit is due partly to its resistance R and partly to its inductance L. The part due to resistance is measured by the product Ra in accordance with Ohm's law, and the part due to inductance is measured by the rate at which La is changing with time, which may be denoted by L$\dot{a}$. The sum of all these differences of potential round the circuit must be zero. Hence we have

$$L\dot{a} + Ra + \frac{q}{C} = 0$$

The current in the circuit at any moment is measured by the rate at which the condenser is gaining or losing charge. Therefore we have with the above notation $\dot{q} = a$, and hence the rate at which a is changing is the rate at which $\dot{q}$ is changing, which may be written $\ddot{q} = \dot{a}$. Substituting these values for a and $\dot{a}$ in the above equation, we have,

$$L\ddot{q} + R\dot{q} + \frac{q}{C} = 0$$

$$\text{or } \ddot{q} + \frac{R}{L}\dot{q} + \frac{1}{LC}q = 0$$

This equation establishes a relation between the charge q in the condenser at any moment, and its rate of change $\dot{q}$, and its rate of rate of change $\ddot{q}$, and it is called a *differential equation*. In all oscillatory circuits which present themselves in radiotelegraphy the high frequency resistance R measured in ohms, the inductance L measured in henrys and the capacity C measured in farads have such values that the second term $\frac{R}{L}\dot{q}$ may be neglected in comparison with $\frac{1}{LC}q$.

Accordingly, we may in general neglect the second term and write the equation for the circuit in the simple form

$$\ddot{q} + \frac{1}{LC}q = 0$$

This equation is the mathematical expression for an oscillatory motion. To find a solution of it applicable to the present case consider the following theorem:—

Let a point P move uniformly in a circle round the centre O

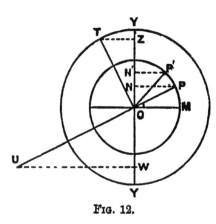

FIG. 12.

(see Fig. 12) in a time T seconds, and let it describe the arc MP in a time t. Then the angle MOP is equal to $\frac{t}{T}$ of four right angles, or to $2\pi\frac{t}{T}$, which we shall write in the form MOP $= pt$, where $p = \frac{2\pi}{T}$. Project OP on the vertical line through O, and then ON = OP sin MOP. Let ON = q, and OP = Q, then we have

$$q = Q \sin pt$$

As P moves round the circle the point N moves backwards and forwards along YY. It is required to find the velocity and the acceleration of the point N. Let P move to P′, and therefore N to N′ in a small interval of time δt. Then the angle POP′ $= p\delta t$ and the arc PP′ $= pQ\delta t$, and if we write δq for NN′ it is easy to see that

$$\frac{\delta q}{\delta t} = pQ \cos pt$$

But $\frac{\delta q}{\delta t} = \dot{q}$ or the time rate of change of ON or q. If, then, we draw a line OT at right angles to OP and make OT $= p$. OP $= pQ$, it is easy to see that its projection OZ on OY is equal to $pQ \cos pt$, and hence the length of OZ is a measure of the velocity of the point N or of $\dot{q}$. If we repeat this process, and draw a line OU at right angles to OT and p times as long or equal to p^2Q, then its projection OW on OY is equal to $p^2Q \sin pt$, and by similar reasoning it can be shown that the length OW is a measure of the acceleration of N or of $\ddot{q}$. If, then, the point N executes a simple periodic motion to and fro along YY, so that its displacement at any instant is ON $= q$, its acceleration $\ddot{q}$ at that instant is equal to $-p^2$ times ON, or $\ddot{q} = -p^2q$.

Returning then to our electrical oscillation, we may make the supposition that the charge q of the condenser varies periodically from Q to zero, and from zero to $-$ Q in accordance with a simple periodic law, so that $q =$ Q sin pt where $p = 2\pi$-times the periodic time T. We have already shown that for the oscillatory electric circuit

$$\ddot{q} + \frac{1}{LC}q = 0$$

Also that for any harmonic motion

$$\ddot{q} + p^2q = 0$$

It follows that

$$p = \frac{1}{\sqrt{LC}}$$

and since $p = \dfrac{2\pi}{T}$, we have for the periodic time T the expression

$$T = 2\pi\sqrt{LC}$$

The electric oscillations, therefore, are executed with a frequency $n = \dfrac{1}{T}$, and

$$n = \frac{1}{2\pi\sqrt{LC}}$$

These two last formulæ accordingly give us expressions for the time-period of the oscillations and their frequency, where T is measured in seconds, L in henrys, and C in farads.

As, however, the capacity is most conveniently measured in microfarads and the inductance in microhenrys, the time-period will be expressed in microseconds, or in millionths of a second. This time-period is called the *natural time-period of oscillation* of the circuit. Just as every pendulum has its own natural time-period of oscillation, in which it executes vibrations when disturbed and left to itself, so every oscillatory electric circuit consisting of a capacity and inductance in series with each other has its own natural time-period in which an electric charge given to it oscillates if disturbed and left to itself.

We may put the formula for the natural frequency of an oscillatory circuit in a more convenient form for calculation. Let L *cms.* stand for the inductance of the circuit expressed in absolute electromagnetic units, that is in centimetres. This is equal to the inductance in microhenrys multiplied by 1000. Let C *mfds.* stand for the capacity measured in microfarads. Then it is easy to show that

$$n = \frac{5 \cdot 033 \times 10^6}{\sqrt{C \; mfds. . L \; cms.}}$$

The number 5·033 is equal to $\dfrac{\sqrt{1000}}{2\pi}$

Thus, suppose that the circuit consists of a Leyden jar having a capacity of $\frac{1}{300}$th of a microfarad, and the inductive circuit has an inductance of 30 microhenrys or 30,000 centimetres. Then the quantity $\sqrt{C \; mfds. . L \; cms.}$ has a value of 10. This last quantity is called the *oscillation constant* of the circuit. The natural frequency of such a circuit is then half a million, and the natural time-period of oscillation is two microseconds. When two oscillatory circuits have the same natural time-period they are called *syntonic circuits*. This is the case when the product of the capacity and the inductance of the two circuits is the same, although the individual values may be very different.

7. **Electric Resonance.**—If an oscillatory circuit has a periodic or alternating electromotive force set up in it and if the frequency of this E.M.F. agrees with the natural frequency of the circuit, then an immensely greater current will be produced than if the periods do not agree. This increase ·in the amplitude of the alternating current created in the circuit by exactly syntonising the frequency of the impressed E.M.F. with the natural frequency is said to be due to *electric resonance*. The term is borrowed from acoustics. It is a familiar fact that we can set up a very considerable amplitude of vibration in a pendulum by administering to it small blows, or even puffs of air, provided that these are timed to agree exactly with the natural period of the pendulum. In the same manner oscillations of great amplitude can be created in a heavy elastic beam supported at both ends by very gentle blows given at the right intervals in the centre.

Supposing that an electric condenser has its plates connected by a wire, part of which forms a coil, and that over this last coil we wind another wire in which a high frequency current can be set up. Then alternating currents in this coil, called the primary coil, generate by Faradaic induction a secondary electromotive force in the other coil, and therefore a secondary current. The secondary current will be small unless the two circuits are syntonised, but it increases very rapidly as the circuits are brought into syntony with each other. This is called tuning the circuits. Two such overlaid or neighbouring coils constitute an *oscillation transformer*, and the two circuits are said to be *coupled* together. If the primary and secondary coils are very close together, or intertwined, the circuits are said to be *closely coupled*. If they are far apart, they are said to be *loosely coupled*.

If both circuits possess inductance and capacity, they are called *coupled oscillatory circuits*, and they can be tuned by varying the capacity or inductance of one or both until the oscillation constants of each circuit are either equal or in exact integer ratio to each other.

If oscillations are set up in a circuit which are in agreement with its natural frequency, they are called *free oscillations*. If, however, oscillations are maintained which have a frequency different from the natural frequency of circuit, they are called *forced oscillations*.

Suppose, then, that we form an oscillatory circuit consisting of a condenser, C, the capacity of which can be varied gradually, and an inductance coil, P, which forms one coil of an oscillation transformer, and insert in the circuit some instrument, such as a hot wire ammeter, A, adapted for measuring high frequency currents (see Fig. 13). Let the other circuits of the oscillation transformer have undamped oscillations set up in it of any frequency. If then we take observations of the reading of the ammeter, beginning with a very small capacity in the circuit, and steadily increase the capacity, we shall find the current as read on the ammeter increases with the capacity up to a certain capacity, and after that decreases again. If we plot these ammeter readings as ordinates of a curve drawn to capacity values as abscissæ, we shall obtain a curve rising up to a peak more or less sharply, corresponding to the maximum observed value of the current, and to a certain value of the capacity (see Fig. 14). This maximum current is called the *resonance current*, and the value of the

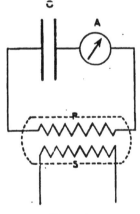

Fig. 13.

natural frequency of the circuit corresponding to the value of the capacity which produces this maximum current is called the *resonance frequency*, whilst the curve itself so drawn is called a *resonance curve*. Such resonance curves can be used, as we shall show in Chapter VIII., for determining the logarithmic decrement or damping of the oscillatory circuit for which they are described.

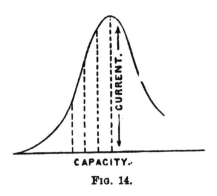

Fig. 14.

If the resonance curve is very sharply peaked, it shows that a very little want of tuning of the oscillatory circuit to the exact frequency of the impressed electromotive force has a very great effect in reducing the current induced in the oscillation circuit. This implies that exact tuning has a great effect in exalting the current. This again shows that the damping or sources of energy loss in that oscillation circuit are small, and hence the logarithmic decrement is small. On the other hand, a resonance curve with a rounded summit, as in Fig. 15, implies that the departure from absolute syntony has not much effect in reducing the current, or, *vice versâ*, that exact tuning has not much effect in exalting it, and this means that the circuit in question has a large decrement.

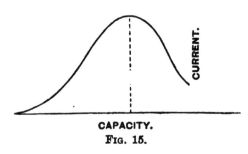

CAPACITY.
Fig. 15.

Undersomecircumstances the resonance curve is a double-humped curve, as in Fig. 16, which indicates that oscillations of two different frequencies exist in the circuit. Let there be two oscillation circuits, each consisting of a condenser and an inductive circuit. Let these circuits be coupled inductively and closely together by overlaying one circuit on the other, so that oscillations set up

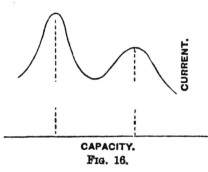

CAPACITY.
Fig. 16.

in one circuit induce oscillations in the other, and let the two circuits be syntonised. Then, if free oscillations are excited in one circuit, the result will be to create in both circuits a complex oscillation, which is alternately of greater and of less frequency than the natural frequency of either circuit taken alone. The effect, however, is capable of simple explanation by a mechanical analogy. Let a string be fastened loosely across a room, and from it let two rods be hung like pendulums placed a little way apart. Draw one rod on one side and let it go, setting it in vibration like a pendulum in a plane at right angles to the line of the sustaining string. It will at once begin to set the other pendulum in vibration, because the first pendulum in its vibrations imparts little jerks to the string, and so administers impulses which set the second pendulum in vibration. It will be found that as the second pendulum begins to take up the vibrations the first one comes to rest. The reason is obvious. By the Third Law of Motion, action and reaction are equal and opposite, and the first pendulum cannot accelerate the second without retarding itself. When the first pendulum has come to rest the second one is in full swing. The process then repeats itself, and the energy is gradually handed back from the second to the first.

If we consider the manner at which each pendulum is swinging, it is not difficult to see that the amplitude or extreme range of motion of each pendulum periodically increases and decreases. Such a motion can be resolved into the sum of two simple periodic motions of slightly different frequencies. Precisely the same thing happens in the case of the two coupled electric circuits. The oscillations in the primary can only create secondary oscillations by generating an electromotive force in the secondary circuit, and a counter-electromotive force in their own circuit. This results from the insertion and withdrawal of lines of magnetic flux into and from the circuits, and the effective inductance of the primary circuit is changed by the presence of the neighbouring secondary circuit. Accordingly, the two circuits act and react on each other like the two pendulums above mentioned, and create oscillations which periodically fluctuate in amplitude, and they are therefore resolvable into oscillations of two different frequencies. If we place near to either the primary or the secondary circuit another or tertiary circuit, which contains a variable capacity and some means for indicating the current strength in it, we can, by altering the capacity in this last circuit, tune it to syntonise with either frequency, and if we plot a resonance curve by continuously varying the capacity in the tertiary circuit, we find it to be a double-humped curve as in Fig. 16, where the two maximum ordinates correspond to the two separate frequencies of the oscillations set up in the secondary circuit. The effect is exactly analogous to that known in music as the *beats* produced by two organ pipes slightly out of tune with each other. If two such pipes are sounded together, we hear a sound which waxes and wanes periodically. This is because the sound waves emitted, having slightly different wave lengths at intervals, exalt each other, and at intermediate intervals partly nullify each other. Hence any motion which is periodic, and has also a periodic variation in amplitude, can be resolved into two simple harmonic oscillations of two slightly different frequencies or wave lengths. We shall return to this matter again in a later chapter in considering the inductively coupled transmitter used in radiotelegraphy, and also the cymometer for measuring electric wave length.

CHAPTER II

DAMPED ELECTRIC OSCILLATIONS

1. The Production of Damped Oscillations by Condenser Discharges.—If a condenser such as a Leyden jar is electrically charged and then discharged through a wire having a high resistance, the flow or movement of electricity set up in the discharge circuit is always in the same direction. The current begins by being zero at the instant of completing the circuit, rises to a maximum very quickly, and then gradually falls in strength to zero again. The variation of the current during discharge may be represented by the ordinates of a curve as in Fig. 1. This form of discharge is called the *dead-beat* or *non-oscillatory* discharge. It takes place when the resistance R of the discharge circuit is greater than the value of $2\sqrt{\dfrac{L}{C}}$, where C

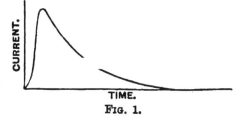

TIME.

FIG. 1.

is the capacity of the condenser measured in farads, L the inductance of the circuit in henrys, and R the resistance of the circuit in ohms.

Thus suppose that a Leyden jar having a capacity of 0·01 of a microfarad or 0.01×10^6 of a farad is discharged through a circuit having an inductance of 1 microhenry. Then, in these units, the quantity $2\sqrt{\dfrac{L}{C}}$ is equal to 20, and hence if the circuit has a resistance of more than 20 ohms the discharge will be dead beat or unidirectional in type.

If, however, the resistance is much less, say 1 ohm, then the discharge will not be unidirectional, but will be an *oscillatory* discharge, electricity moving backwards and forwards in the discharge circuit with gradually decreasing amplitude, in the manner already explained in the previous chapter, so that the discharge current is represented by the ordinates of a decrescent wavy curve, as in Fig. 2, constituting a train of damped oscillations.

A mechanical illustration of these two modes of discharge is as follows:

Imagine two metal vessels, one of which is exhausted of its air and the other contains air under pressure. The difference of the pressure of air in the two vessels corresponds to the difference of potential between the two plates of the electric condenser. Let the above-named vessels be connected by a long, narrow pipe

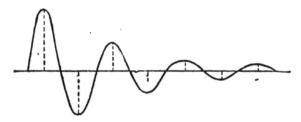

FIG. 2.

which offers considerable obstruction to the movement of air through it. Then when the connection is made by opening a tap in the pipe, the air pressure will sink in one vessel and rise in the other, but the motion of the air in the pipe will always be in the same direction. This corresponds to the dead-beat electric discharge. If, however, we suppose the vessels to be connected by a short, wide pipe offering but little obstruction to the motion of air through it, then, if the connection is suddenly established by opening a valve, the air pressure in the two vessels will only be equalised after a series of rushes of air to and fro in the pipe. The air first flows through the pipe in one direction, and then in virtue of its inertia overshoots the mark and moves back again, and this action is then repeated, each of the air oscillations being less than the previous one, until they subside and leave the two vessels at last at equal pressure.

To create trains of damped electric oscillations we must therefore charge some form of electric condenser and discharge it suddenly through an inductive circuit of low resistance, and repeat the process again and again. The simplest way of doing this is by means of an induction coil having its secondary terminals connected to two metal balls separated by a slight air gap, these balls being also connected to a condenser and low resistance inductance coil joined in series, as in Fig. 3.

The induction coil I has its primary circuit supplied with current from a primary or secondary battery or from a continuous current dynamo, and in this circuit is an automatic interrupter which continually opens and closes the battery circuit. At each interruption an electromotive force is set up in the secondary circuit, and the condenser C becomes charged with positive electricity on one side and negative on the other. When the difference of potential of the two plates reaches a limit fixed by the length of the gap between the two spark balls, a discharge of the condenser takes place through the inductive resistance L and across the spark gap S, and if the gap is not too long the conditions are fulfilled for

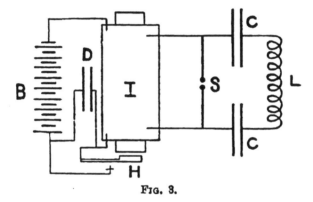

Fig. 3.

the production of the oscillatory form of discharge of the condenser which accordingly takes place. Several such oscillatory discharges may take place during the rise and fall of the electromotive force created in the secondary circuit at each interruption of the primary current. Hence, when the coil is set in operation a bright crackling spark appears in the spark gap, and the experienced ear can decide from the sound whether this spark is accompanied by electric oscillations or not.

This apparatus is called the *spark apparatus* for the production of damped oscillations. A perspective view of it is shown in Fig. 4. The oscillations die away in amplitude during the train because the energy of the original condenser charge is being dissipated by resistance or in other ways, part of the dissipative resistance residing in the spark itself. We shall then consider each of the element of this apparatus in succession.

2. **Induction Coils for Radiotelegraphy.**—We shall assume that the reader is acquainted with the general construction and mode of operation of an induction or spark coil. For the purposes of radiotelegraphy a type of induction coil frequently employed is one

called a 10-inch coil, meaning an induction coil which can produce a spark 10 inches long between points or small balls attached to the ends of its secondary circuit. Such a coil consists of a primary circuit of thick wire, generally No. 12 S.W.G. in size, wound in 300 or 400 turns on an iron core composed of a bundle of well annealed soft Swedish iron wires, each not more than No. 22 S.W.G. in thickness, the bundle being 2 inches in diameter and about 18 inches long. In some cases makers provide several

FIG. 4.

primary cores or wind the primary coil in several distinct layers, the ends being brought out so as to combine them in different ways in parallel or in series. This, however, is not so necessary in radiotelegraphy as in Röntgen ray work.

The resistance of the primary of a 10-inch coil made as above would be about 0·3 of an ohm, and its inductance 0·02 of a henry. This primary coil is enclosed in an ebonite tube with very thick walls, at least ¼ of an inch thick, and this tube must be very carefully made and perfectly free from flaws. The tube should be 2 inches longer than the iron wire bundle and closed by ebonite caps at the ends, the ends of the primary wire being brought out through holes in one of the caps. This tube carries two thick ebonite cheeks between which the secondary coil is wound. This last consists of a very considerable length (10 to 17 miles) of No. 34 or No. 36 S.W.G. copper wire, double covered with white silk. The secondary circuit is wound in a very large number of flat coils or sections, several hundred such coils being sometimes employed. These are prepared by winding the silk-covered copper wire between paper discs in a flat spiral, as a sailor winds up a spare rope. These coils are then slipped on to the ebonite tube enclosing the primary coil, and the ends of the coils are then jointed together. To enable this to be done, the coil sections are wound in double flat layers with a disc of paraffined paper between, the beginning and end of the wire thus being at the outside, and the two layers so wound that the windings follow on in the same direction. There is then no difficulty in making the joints between the various flat coils composing the secondary circuit. Mr. Leslie Miller has, however, invented an ingenious machine for winding the flat disc coils consecutively with no joints between them at all, as shown in Fig. 5, in which, however, the discs are

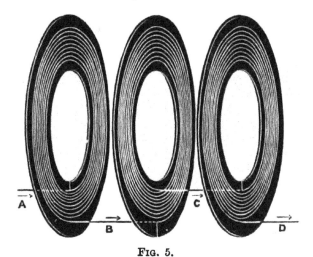

FIG. 5.

shown widely separated for the sake of clearness. The object of this mode of winding is to secure that no two points on the secondary wire which are at great differences of potential come near to each other. The whole of the very numerous flat coils forming the secondary circuit are compressed together between the two thick ebonite cheeks, and it is usual to immerse the whole finished secondary coil in very hot melted paraffin wax to exclude air and insulate it thoroughly. The coil so made is finished by enclosing in an outer sheath of thin ebonite and mounting it on a box baseboard on supports of ebonite. The ends of the secondary wire are brought to two terminals carried on ebonite pillars and provided with adjustable spark points (see Fig. 6). Two other adjuncts are then necessary. In the first place some means has to be provided for rapidly interrupting or reversing the primary current, since it is only by so doing that we can create electromotive force in the secondary circuit. This appliance is called an *interrupter*. Also it is necessary, as first shown by Fizeau, to place a condenser of a certain capacity across the points between which interruption of the primary circuit occurs. This is called the *primary condenser*.

In reference to the construction of induction coils for radiotelegraphy, it should be noticed that the value of a coil is not to be judged simply by the length of spark it can give between the ends of the secondary wire when the primary circuit is interrupted,

FIG. 6.

[Reproduced by permission of Messrs. Newton & Co.

but by the spark length which can be obtained when a condenser of a certain capacity, say 0·1 microfarad, is connected across the ends of the secondary circuit.

At each make and break of the primary circuit an electromotive force is created in the secondary circuit, the magnitude of which depends upon the rate at which the magnetic flux created in the core is linked or unlinked with the turns of the secondary circuit. This flux grows up when the primary current starts more slowly than it decays as the primary current stops, and hence the secondary electromotive force at the "make" of the primary circuit is very small compared with that at the "break." The latter may be several hundred times greater than the former. Hence, if the primary circuit is rapidly made and broken, an intermittent electromotive force appears in the secondary circuit which is practically always in the same direction, only that secondary electromotive force due to the stoppage of the primary circuit being of any importance. This electromotive force lasts a very short time at and during the decay of the magnetism in the core. If we charge a condenser having a capacity of C farads through a non-inductive resistance of R ohms at a source of constant electromotive force of V volts, then it can be shown that the condenser will not be charged to the voltage V unless the electromotive force is kept applied to the condenser for a time at least equal to 5 times CR seconds. Thus, if we have a condenser of 1 microfarad and try to charge it through a resistance of 1 megohm (1 million ohms) at a pair of terminals giving 100 volts continuous supply, the condenser will not be charged to 100 volts in much less than 5 seconds. At the end of 1 second it will only be charged to 63 volts, and at the end of 2 seconds to.86 volts, and at the end of 3 seconds to 95 volts, as shown by the ordinates of the curve in Fig. 7. Hence, although a very large electromotive force makes its appearance in the secondary at each break

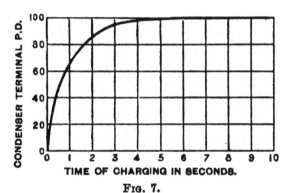

FIG. 7.

of the primary circuit, since it has to act through the high resistance of the secondary coil, amounting to several thousand ohms, and lasts only a very small fraction of a second, it is only able to charge a condenser connected between the secondary terminals to a small fraction of the maximum voltage which would exist in the secondary circuit if the condenser were removed.

Suppose, then, that we attach to the ends of the secondary circuit of a large induction coil a pair of metal balls adjustable as to distance, and insert a proper interrupter and primary condenser in the primary circuit. On interrupting the primary current we obtain a torrent of thin blue irregular lightning-like sparks between the secondary balls, and by extending the distance between these balls we can produce a spark, say, of a maximum length of 10 inches, and this implies that the secondary circuit is the seat of intermittent electromotive forces rising as high as 500,000 volts momentarily. If then we join the outside and inside coatings of a large Leyden jar having, say, a capacity of 0·01 microfarad to

these secondary terminals, we find that we can get no spark until
the balls are brought within a few millimetres (5 to 10) of each
other. The condenser cannot therefore be charged to more than
20,000 or 30,000 volts. Any increase in the resistance of the
secondary circuit or of the capacity of the condenser decreases the
effective spark length and charging voltage. Accordingly, it is
desirable to keep the resistance of the secondary circuit as low as
possible by employing a rather thick secondary wire. This, however, implies increased cost of manufacture and increased bulk of
coil, unless the number of turns of the secondary circuit is
correspondingly decreased. It is found, however, that a certain
compromise can be effected, and that a coil constructed, say, with
a secondary circuit of No. 30 S.W.G. wire will enable a certain
fixed charging voltage or spark length to be obtained on a condenser of larger capacity than by means of a coil wound with the
same weight of secondary wire of a smaller diameter, such as
No. 36 S.W.G.

Coils with rather thick secondary wires are called "intensified
induction coils" by Messrs. W. Watson & Sons, and by Messrs.
Newton & Co. "heavy discharge coils." They have advantages
as above described for radiotelegraphic purposes. The same
result can, however, be achieved by coupling together in
parallel the similar terminals of the secondary circuits of two or
more identical coils having fine wire secondary circuits and joining
their primary circuits in series with each other, and with one
interrupter and primary condenser. The coils then give their
secondary electromotive forces at the same instant, and they have
the same value, but the effective resistance of the secondary
circuit is reduced. A 10-inch induction coil, as usually made,
takes a mean primary current of 10 amperes, and requires a
terminal voltage of 16 to 20 volts. This current is the
average current, but at certain instants the current actually
taken is much larger. The current is most conveniently supplied from secondary or storage cells. On board ships these
can be charged from the electric lighting circuits, but in lighthouses and lightships it may be necessary to charge these
secondary cells from a number of large primary cells. In this case
suppose a battery of 8 secondary cells has to be charged. This
will require a maximum voltage of 20 volts, and a large number of
primary cells, each having an electromotive force, say, of 1·5 volts,
must be arranged 15 in series, and, say, 6 or more in parallel,
to provide the current necessary to charge the 8 secondary cells.
The secondary cells are then used directly on the coil. The primary
cells used are generally some pattern of low resistance dry cell.

It will be seen, therefore, that the induction coil described requires a power supply of about 150 watts, or, say, 0·2 of a horse-power. This is sufficient for radiotelegraphic work up to 100 miles, but for greater distances more power is required, and this is obtained by the use of alternating current transformers. Before describing these high power arrangements it will be convenient to complete the description of the induction coil by referring to the various forms of current interrupter used.

3. **Interrupters for Induction Coils.**—As already explained, the use of this appliance is to interrupt rapidly and very suddenly a continuous current flowing from the battery through the primary circuit of an induction coil, so as to create a momentary electro-motive force in the secondary circuit. The magnitude of this secondary E.M.F. depends, other things being equal, upon the suddenness with which the primary current is arrested.

In the early forms of induction coil, a hand-worked mechanical circuit breaker was employed, but it was not long before automatic interrupters were invented, and very much ingenuity has been expended of late years in devising varieties of interrupter for this purpose.

At the present time we may divide known forms into three broad classes—

(i) Hammer or plunger interrupters.

(ii) Mercury jet or turbine interrupters.

(iii) Electrolytic interrupters.

The hammer interrupter in its simplest form consists of a cylindrical block of soft iron which is mounted on the top of a stiff flat brass spring, the lower end of the spring being attached to the baseboard of the coil. The iron block is fixed opposite to the end of the soft iron core of the coil, and when the latter is magnetised it attracts the iron block or hammer head. The back of this block is furnished with a thick platinum stud, which has opposite to it a similar stud carried on the end of a screw which is tapped into a fixed pillar also mounted on the baseboard. In normal adjustment when the primary current is not flowing, the two platinum studs just make contact, and the primary current flows into the coil through this contact. The core is then magnetised and attracts the hammer head, which separates the platinum contact studs and interrupts the circuit. To make the break more sudden, Mr. Apps added an adjusting screw by means of which the spring carrying the hammer head is forced back so as to make a stronger contact against the contact stud (see Fig. 8).

In this form it is now known as an Apps break. With such a break properly adjusted it is possible to vary the number of interruptions of the primary circuit from about 10 to 50 per second.

The platinum contact studs require some care to keep them in order. They become rough and are burnt away by the spark which takes place at the point of separation, and from time to time require dressing up with a fine file or glass paper. In coils used for radiotelegraphy these platinum studs should be as massive as possible, not less than $\frac{3}{16}$ inch in diameter and $\frac{1}{4}$ inch in length, and larger if possible. Platinum costs now about £4 or more per ounce, and being a very dense metal, the expense of large contact studs is not inconsiderable. Although this hammer break is irregular in action and has notable defects, yet it is simple and easy to keep in order, and hence has been much used on radiotelegraphic coils, taking a primary current of not more than 10 amperes, at a low voltage, say, 12 to 18 volts.

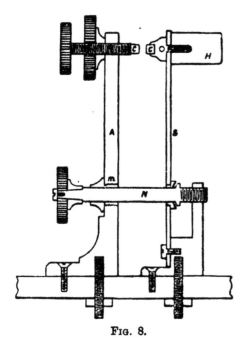

Fig. 8.

It is, however, not suitable for larger currents or higher voltages.

The interrupter is, however, incomplete without the addition of a primary condenser. This last consists of sheets of tinfoil separated by somewhat larger sheets of paraffined paper. The condenser is built up as follows :—

Sheets of thin but good linen paper are dried and immersed in melted paraffin wax and then lifted out and allowed to cool. Sheets of tinfoil rather less in dimensions than the paper are provided. These tinfoil sheets are separated by double or treble sheets of the paraffined paper, and each alternate sheet of tinfoil is displaced so as to project beyond the paper to one side or the other. The alternate sheets of tinfoil are then pinched together and the mass strongly compressed. It then forms a condenser of which all one set (say the even numbers) of tinfoil sheets form one plate or coating, and all the other sets (say the odd numbers) form the other coating, whilst the paraffined paper is the dielectric.

This condenser is generally placed in the box base which carries the induction coil, and the two coatings are connected respectively to the spring carrying the hammer head and to the upright carrying the contact screw.

The usual connections of the complete induction coil are shown in Fig. 9. In this diagram P and N are the terminals to which the battery is attached. T is the platinum contact and H the

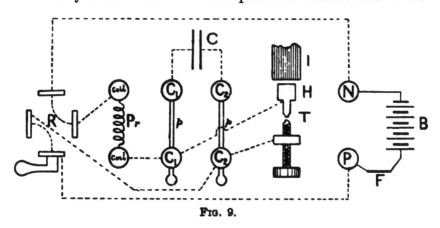

FIG. 9.

hammer break. I is the iron core of the coil. C is the primary condenser. Pr is the primary of the coil, and R is the current reverser.

The primary condenser of a 10-inch induction coil would generally have a capacity of about 1 microfarad, which might necessitate 100 square feet of tinfoil, reckoning both surfaces; in other words, would consist of 50 sheets of tinfoil, each 12 inches by 12 inches in area, separated by double sheets of paraffined paper. The exact capacity of the condenser does not seem to be very important, provided it is not too small; the coil gives a very small secondary spark without it, and it therefore is an important element in the coil construction. Lord Rayleigh found that if the primary circuit is interrupted with extreme suddenness, for instance, if severed by a bullet from a gun, the secondary spark length is not increased by the addition of the condenser. With the ordinary break speed the primary condenser acts to annul more quickly the magnetisation of the coil at each break of the primary circuit, and it appears to do this by destroying the electric arc which tends to form across the contact studs as they separate.

A second type of circuit breaker is represented by the mercury jet interrupter (see Fig. 10). It consists of a cylindrical block of steel which is carried on a shaft and has its lower end immersed

E

in mercury.   The block has a vertical hole drilled part of the way through it and a lateral hole meets the vertical one.   If the block is made to revolve rapidly, the mercury is sucked up the vertical hole and thrown out of the lateral one in the form of a jet.   If

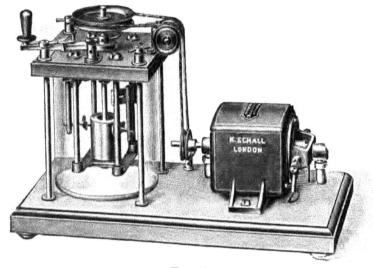

FIG 10.

this jet impinges on a copper plate or plates placed in the right position, then when the jet is playing on the plate there is an electric circuit through the jet from the mass of mercury beneath the centrifugal pump to the target-plate, but when the jet squirts aside of this plate the circuit is interrupted.

To prevent this jet of mercury from being rapidly oxidised, the whole process must take place in some non-oxidising medium, such as oil, alcohol, or, better still, coal gas.

In one practical form the mercury jet interrupter consists of a glass vessel in the bottom part of which is the reservoir of mercury. The cover of this vessel carries the steel shaft on the lower end of which is the pierced steel block constituting the centrifugal pump.   This shaft is rapidly rotated by a small electric motor. The lid also carries an insulated terminal to which is attached the copper plate which forms the target against which the revolving jet of mercury issuing from the pump impinges during a portion of its revolution.   The circuit is interrupted suddenly at each revolution between the terminal attached to the target plate and the shaft of the pump at the instant when the jet ceases to play on the plate.   In some interrupters the mercury is covered with paraffin oil to a depth sufficient to include the jet. This oil, however, prevents the mercury globules into which the jet breaks up from cohering again, and before very long the whole mass of mercury is reduced to the condition of a fine black sludge. It can then only be recovered to the condition of bright metallic mercury by a troublesome process of treating it with strong sulphuric acid and washing away the acid with water. It has been found that when using absolute alcohol instead of oil as the insulator the mercury keeps coherent for a longer time, but there are disadvantages in using such an inflammable and volatile liquid as alcohol. Hence, although the mercury jet interrupter in oil or alcohol has been much used as a break for radiotelegraphic coils, and is capable of giving a more regular and rapid series of interruptions than the hammer break, it is far from being perfect or very convenient in use.

A great improvement, however, was made by M. Béclère of Paris in this break by substituting an atmosphere of coal gas for the oil or alcohol. In this improved form it is known as the moto-magnetic or coal gas mercury break. By replacing the liquid by a gas much less power is required to drive the break. Hence an electric motor is not necessary when the break is continuously used. The break may then consist of a metal chamber in the bottom of which is a small quantity (generally less than 1 pound) of mercury. Into this the lower part of a cylindrical block of steel dips, the block having a channel cut in it, so that when the block revolves the mercury is sucked up the channel and squirted out in the form of a revolving jet, which intermittently impinges upon a metal plate or plates insulated from the containing box. This centrifugal pump is caused to revolve by having fixed on its shaft a Z-shaped piece of soft iron. See Fig. 11, which shows a type of this break made by Messrs. W. Watson & Sons.

FIG. 11.

The interrupter is mounted on the induction coil base so that this Z-shaped piece comes opposite to one end of the iron core.

If the core is periodically magnetised and demagnetised, it exerts attractive impulses on the iron armature and maintains it in motion. The break is so connected in the primary coil circuit that the current flowing through it passes through the jet of mercury and then magnetises the core of the coil. This magnetism is made to attract the Z-shaped iron and pull it round into a position in which the jet no longer impinges on the target plate. The circuit is then interrupted and the coil core loses its magnetism. The pump, however, continues to revolve by its own inertia and immediately closes the circuit again and so repeats the process. To start the break an initial twist must be given to it by hand and then after that the intermittent magnetic impulses of the coil core maintain it in motion. To prevent oxidation of the mercury the chamber is kept full of coal gas. This is supplied from a small steel bottle, and as the gas is not used up but only exhausted very slowly by leakage a cylinder holding 5 cubic feet of gas will last for months. This interrupter has been found to work very well for Röntgen ray coils where continuous operation of the coil is necessary. For radiotelegraphic work it is necessary, however, to have the interrupter driven by a small separate electric motor worked off the same battery which supplies the coil, for in radio-telegraphic work the coil is not continuously in operation, but is started and stopped intermittently, and the interrupter must there-fore start into action at once or be kept in action. Any ordinary oil or alcohol mercury turbine interrupter can be improved and converted into a coal gas mercury break by making the vessel in which the turbine revolves gas tight. It is then only necessary to expel the air from this chamber by coal gas drawn from a small steel bottle or gas bag which is kept connected with the vessel so as to supply the small gas leakage and prevent air diffusing back into the chamber.

The third type of interrupter is the electrolytic break. If a large lead plate is immersed in dilute sulphuric acid containing 20 per cent. of acid, and if a second electrode of very small surface is immersed, consisting of a short length of stout platinum wire, projecting for about $\frac{1}{8}$ of an inch out of the end of a tight fitting glass or porcelain tube, then such an electrolytic cell interposed in the primary circuit of the induction coil will interrupt the current, provided that the electrode of small surface is the positive electrode or anode (see Fig. 12).

This form of interrupter was invented by Dr. Wehnelt in 1899, and excited great interest at the time of its introduction. If a continuous electromotive force is applied to a circuit consisting of a Wehnelt electrolytic cell made as above described, and an inductive resistance, the current is not continuous, but is intermittent, several hundred interruptions occurring per second. The action essentially depends upon the presence of inductance in the circuit, and if the inductance of the primary coil alone is not sufficient, an extra inductance coil must be added in series with

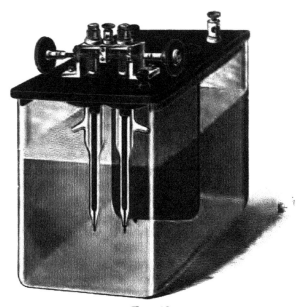

Fig. 12.

[*Reproduced by permission of A. C. Cossor.*]

it. No primary condenser is necessary. When such a Wehnelt interrupter is attached to the primary circuit of an induction coil, the secondary spark is of a different character to that given by a hammer break. It is no longer a long, thin, irregular, lightning-like spark, but becomes shorter and more like a flame or alternating current arc, owing to the large number of the interruptions per second.

A modification of the Wehnelt break, introduced by Caldwell and by Campbell Swinton, consists of a porcelain test tube, with a small hole in the bottom. This is immersed in a larger vessel containing dilute sulphuric acid (20% acid). A large lead plate is immersed in the outer vessel, and a smaller lead plate in the test tube, and it is found that when this electrolytic cell is placed in series with an inductive resistance, a continuous electromotive force or battery produces a rapidly intermittent current in the circuit. The frequency and strength of this current can be regulated by more or less closing the small hole in the test tube by the conical end of a glass rod inserted into it.

Although the electrolytic break has been much used in connection with Röntgen ray and for certain electro-medical work, its advantages in connection with radiotelegraphy are not nearly so great.   The reason for this is as follows :—

If a condenser has its plates connected to the secondary terminals of an induction coil, and if an electromotive force begins to arise in the secondary circuit, the condenser begins to receive a charge.   If the secondary terminals end in spark balls placed a certain distance apart, the potential difference of the condenser plates will increase up to a limit determined by the interval between these spark balls.   At that moment a discharge takes place, and the condenser discharges with oscillations across the spark gap.   If at the same time the supply of electricity from the coil is sufficiently rapid to cause a true electric arc to take place between these balls, the result is to reduce their difference of potential very considerably, and it will not increase again to a high value until the arc stops or is extinguished.   The discharge which takes place is therefore a compound effect.   It consists of an arc discharge superimposed on a condenser oscillatory discharge, and before the latter can be repeated the arc must be extinguished in order that the difference of potential between the balls may rise again to a value sufficient to charge the condenser with a sensible amount of energy.   It is found that if the source of potential difference can supply a large current, then special means have to be adopted to repress this arc.   The discharge between the spark balls should, in fact, consist of electricity which has come out of the condenser, and not directly out of the electric generator employed to charge the condenser.   We shall describe presently methods which are employed to repress this arcing.   Meanwhile it may be mentioned that one method by which it can be achieved is by blowing a jet of air upon the spark gap.   This prevents the arc, but does not hinder the true condenser oscillatory spark discharge from passing across the gap.

4. **Alternators and Transformers for Radiotelegraphy.**—When electric oscillations of large power have to be created, the induction coil and interruptor operating with the current from a battery are not suitable for the purpose, but alternating currents of high potential generated by alternators and transformers are employed.

The most convenient form of alternating current generator for radiotelegraphy is a form of dynamo called a revolving field *alternator*.   This consists of an iron frame or carcase built up

of thin sheets of iron stamped out in the form of a ring with
inward projecting teeth. Around these teeth are wound coils of
insulated wire which are connected up in series with each other.
In the centre of the ring is placed a star-shaped electromagnet,
with an even number of pole pieces, each surrounded by a coil
of wire. These coils are joined in series, and the ends brought to
two insulated rings on the shaft of the magnet. By means of
springs or brushes pressed against these rings a continuous current
from a secondary battery is passed through the magnet coils, and
magnetises them, so that alternate pole pieces are north poles
and south poles. This so-called field magnet is set in rotation by
an engine by means of a pulley and belt, or else by a direct
coupled engine, and as it revolves the lines of magnetic flux proceeding from its poles sweep through the fixed coils attached
to the inner side of the ring, which are called the armature coils.
In these coils alternating electromotive forces are created, and,
since the separate coils are joined in series, these are added
together, and produce a high resultant electromotive force. It is
usual to excite the field magnets by fifty secondary cells or at
a voltage of 100 volts, and to design the alternator to give 2000
volts in root-mean-square or effective value, at a frequency of
300 to 500 periods per second.

If the alternator is a small size, say 5 to 10 kilowatts, or 7 to
14 H.P., it may be driven by a belt and pulley from the flywheel
of an oil or steam engine. In places where coal and water are not
easily procured, an oil engine having two heavy flywheels on its
shaft, from one of which the alternator is driven, is a very convenient source of power. For more permanent or larger installations, where water and coal are available, a steam engine is better,
as the turning moment is more uniform (see Fig. 13).

The current from the alternator at 2000 volts is then passed
into the primary circuit of one or more transformers, and from the
secondary circuits a current at a potential of 20,000 or 30,000
volts, or more, can be drawn.

Transformers for radiotelegraphic purposes must be oil insulated. They consist of a primary circuit and secondary circuit of
insulated copper wire wound over a laminated iron core, the whole
being immersed in an iron vessel under highly insulating oil. The
containing vessel itself should be insulated, and the transformers
generally have their primary or low tension circuits joined in
parallel on the alternator, and their high tension or secondary
circuits may be joined in series.

In working with extra high tension transformers used for
charging condensers, the utmost precautions must be taken to

avoid touching any "live" wire, because such a mistake would in all probability be instantly fatal. Transformers have the power of giving a current far in excess of that which can be furnished by an induction coil, hence the reason for extreme caution in handling them. If such a battery of high tension transformers has its

Fig. 13.

secondary terminals connected to a pair of spark balls, and if these are also joined to the terminals of a battery of condensers C (see Fig. 14), we have an arrangement capable of producing electric oscillations of great power.

Usually the alternator A produces an alternating current at a voltage, say, of 2000 and frequency of 300 to 500. The transformers T transform up this voltage, say, to 20,000, correspondingly reducing the current. The condenser C is thus charged to 20,000 or 30,000 volts or so, and the spark gap S may be 5 to 10 millimetres in length. The condenser then discharges across this spark gap at intervals, producing at each discharge a train of high frequency oscillations, the frequency being determined by the capacity of the condenser and the inductances, P, L, in series with it. It might be thought that since the charging voltage is alternating, or first in one direction and then in the other, that one condenser discharge or train of oscillations would occur at each instant of maximum charging voltage, and therefore that the number of condenser discharges

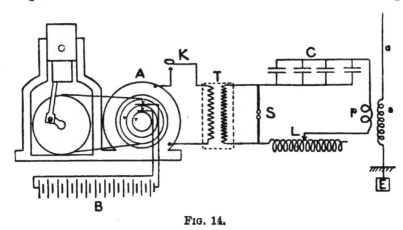

Fig. 14.

per second would be double the frequency of the charging voltage. This, however, is not the case. The number of condenser discharges per second may be much greater or much less than twice the frequency of the alternating current which is supplied to the transformers for the following reasons:—

If the transformers have a very high secondary inductance, or have inductance coils placed in series with their secondary circuits between the transformer terminals and the spark balls, and if the spark balls are set near together, the condenser may become charged to a potential which will enable it to discharge across this gap long before the periodic electromotive force of the transformer has reached its maximum value, as it increases from zero. The moment the discharge happens the condenser empties itself, and *if there is no sensible arc discharge due to the transformer itself* between the spark balls following on this discharge, the still rising electromotive force may again proceed to charge the now discharged condenser, and repeat the process several times during a complete period of the alternator. Hence there may be 3, 4, or more condenser discharges in each half period or alternation of the low frequency electromotive force. This, however, can only happen when the secondary circuit of the transformer is so throttled by inductance as to prevent any true arc discharge from following on after each condenser discharge.

On the other hand, if the inductance of the transformer secondary circuit is not sufficient to prevent this arcing, then there will be no repeated condenser discharge, but only a simple alternating current arc between the spark balls. We can then only obtain oscillations by employing some means to stop the arc without stopping the condenser discharge. This may be done in several ways—

(i.) We may blow a strong jet of air upon the spark gap. This has the effect of blowing out the arc, but it does not prevent the condenser from discharging across the gap.

(ii.) We may enlarge the spark gap to such a length that the normal voltage of the transformer cannot maintain an arc across this distance.

It is then necessary to tune or syntonise the condenser circuit and the secondary circuit of the transformer to the low frequency of the alternator, and when this is done the effect of the resonance will be that at intervals corresponding to, say, 4 or 5 periods such a potential difference will be accumulated between the spark balls that the condenser can discharge across the gap, but the arc due to the transformer electromotive force pure and simple cannot be maintained, because the spark gap is too long.

(iii.) The spark gap may be formed between two surfaces in rapid relative motion, in which case the condenser discharge takes place at the moment of the nearest approach of the surfaces, but the arc is broken by their separation the moment afterwards.

The precise details of these devices are considered in a later chapter, and partly in the next section.

The practised ear soon learns to judge from the sound of the spark whether there is any sensible arc discharge mixed up with the condenser oscillatory discharge. The discharge we desire to produce is that in which the spark between the balls is wholly due to energy which is liberated from the condenser, and not at all to current coming directly out of the transformer. The former is called an *active spark*, because it is the accompaniment of active oscillations; the latter is more or less inactive, and is called an *arc spark*.

5. Spark Discharges and Spark Voltages.—It will be seen that to produce intermittent damped oscillations, as already described, we have to join in series with each other a spark gap, a condenser, and an inductive circuit of some kind, and to employ a source of high electromotive force to charge the condenser, which then discharges at intervals across the spark gap. The damping or logarithmic decrement of the trains of oscillation so produced depends, amongst other things, upon the resistance of this oscillatory circuit, and this, in part at least, comprises the resistance of the spark itself. We have also seen that it is essential to prevent the formation of an electric arc discharge across the spark gap. Hence the nature of the spark gap very much affects the nature of the condenser discharge, and we have to consider the various types of spark discharger employed in radiotelegraphy.

In the case of small power apparatus, including an induction coil as a source of electric supply, it has generally been the custom to take the discharge between brass balls about half an inch or one or two centimetres in diameter, attached to the ends of brass rods which slide through sleeves connected with the induction coil secondary circuit terminals. A rough adjustment of the spark length is then possible, but usually no means is provided for measuring the spark length. With the usual 10-inch coil and a condenser consisting of a battery of Leyden jars the discharge spark may be about 5 to 10 millimetres in length, and makes a rapid series of explosive sounds or continuous crackling noise, but since the quantity of electricity passing is not large the balls themselves do not wear away very rapidly. The chief objection to this form of discharger is the noise it produces. In sending radiotelegraphic signals by the Morse code it is necessary to make a rapid series of oscillatory discharges of long and short duration, the long consisting of 6 or 8 trains of condenser discharges and the short of 2 or 3. These long and short series are arranged according to a certain code of alphabetic signals, as explained later on. To prevent these sound signals from being heard outside the operating room, it is best to enclose the spark balls in an air-tight case with thick walls, by preference a cast-iron vessel, through the walls of which the discharger wires are brought, but insulated by thick ebonite sleeves (see Fig. 15). If, however, the discharge spark

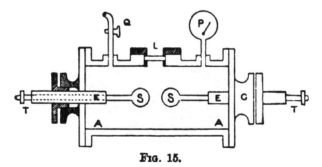

Fig. 15.

takes place in air, it combines some of the nitrogen and oxygen into oxides of nitrogen, which are converted into nitrous and nitric acids, and these deteriorate the insulation of the spark balls. Hence, when operating in a completely closed chamber, it is desirable to replace the air by carbonic dioxide. This can easily be done by means of a supply of the compressed gas kept in a small steel bottle, which is connected with the spark chamber and admitted as required to expel the air.

Even when a complete metallic enclosure is not used some reduction in noise may be obtained by placing the spark balls inside a glass cylinder having very thick walls, which is closed with ebonite or wooden ends, and placing in this cylinder a quantity of quicklime to take up the acid nitrous vapours formed by the spark.

Fessenden found that advantages are obtained by taking the spark between a plate and a point in compressed air contained in an iron drum, and asserted that, when the pressure exceeded 60 lbs. on the square inch, the spark-producer became more efficient as a means of creating electric radiation. The dielectric strength of air or other gases, that is, the potential difference between certain surfaces which is required to produce a spark between them, is approximately proportional to the pressure, when this is greater than one atmosphere. Therefore, the energy given to a condenser connected across a spark gap, which varies as the square of the voltage, must increase roughly as the square of the air pressure for a given spark length. Accordingly, the advantage obtained by increasing the pressure round the spark balls is chiefly due to the fact that we can greatly increase the charging voltage of the condenser without increasing the length of the spark, and therefore also its energy dissipating resistance. The function of the spark gap is to enable the source of electromotive force to charge the condenser, and when this charge reaches a certain value the insulating quality of the air suddenly disappears, and the space instantly becomes conductive, luminous, and highly heated, thus forming an electric spark. The spark is a conductor as long as it lasts. Hence, the two qualities of the air gap between the spark balls with which we are concerned are—

(i.) The dielectric strength or spark voltage; and
(ii.) The spark resistance or energy dissipating power.

The potential difference which is necessary to create a spark discharge between two metal balls depends upon (i.) the size of the balls; (ii.) their distance; (iii.) the nature of the surrounding gas and its pressure; and (iv.) on the temperature of the balls, on the material of which they are made, and on some other conditions, such, for instance, as whether they are exposed to ultra-violet light or the radiation from other sparks. This spark potential difference is generally called the spark voltage, and is stated for a given spark length.

The spark discharge is facilitated or spark voltage lowered by permitting ultra-violet light, *e.g.* the light from an electric arc, the sun, or from electric sparks to fall on the spark balls, especially on the negative ball. The spark discharge is also promoted by heating the electrodes or balls, particularly the negative ball.

In connection with radiotelegraphy we use the spark gap to permit an oscillatory discharge, and it is found that when small quantities of electricity only are in question, the oscillatory form of discharge is promoted by keeping the balls highly polished. For this purpose, well-burnished zinc or brass balls form the best spark surfaces. Accordingly we require to know the potential difference which must exist between two smooth metal balls, say of brass, 1 or 2 cms. in diameter, which will cause a spark to pass between them when their surfaces are at a certain distance apart and the air around them at ordinary or atmospheric pressure.

The following table gives the spark voltage for metal balls 2 cms. in diameter in ordinary air. The distance in millimetres is the distance between the nearest points on the balls.

It should be noted, however, that the figures must not be applied in cases where the balls have diameters differing much from 2 cms., and especially that they are totally inapplicable in the case of points.

TABLE OF SPARK VOLTAGES FOR POLISHED METAL BALLS 2 CMS. IN DIAMETER IN AIR AT 760 MM. PRESSURE.

Spark length in millimetres	Spark voltage.	Spark length in millimetres.	Spark voltage.	Spark length in millimetres.	Spark voltage.
1	4,700	18	44,700	35	61,100
2	8,100	19	46,100	36	61,800
3	11,400	20	47,400	37	62,400
4	14,500	21	48,600	38	63,000
5	17,500	22	49,800	39	63,600
6	20,400	23	51,000	40	64,200
7	23,250	24	52,000	41	64,800
8	26,100	25	53,000	42	65,400
9	28,800	26	54,000	43	66,000
10	31,300	27	54,900	44	66,600
11	33,300	28	55,800	45	67,200
12	35,500	29	56,700	46	67,800
13	37,200	30	57,500	47	68,300
14	38,700	31	58,300	48	68,800
15	40,300	32	59,000	49	69,300
16	41,300	33	59,700	50	69,800
17	43,200	34	60,400	51	70,300

It will be seen from the above table that whilst it requires a potential difference of 4700 volts between the balls to create a spark 1 mm. long, the increase of spark voltage per millimetre continually decreases from about 3000 volts per millimetre for spark lengths of 5 mm. to about 500 volts per millimetre for spark lengths of 50 mm., or so. It is usually said, therefore, that the dielectric strength of air is greater for thin layers of air than for thick ones. This, however, is due to our mode of measuring the spark length, by the distance between the nearest parts of the opposed surfaces, and to the distribution of the electric force in the space. There is reason to believe, as Dr. A. Russell has shown, that the true dielectric strength of air is a fixed constant, equal to 3800 or 3900 volts per millimetre.

In the case of sparks conveying small quantities of electricity the balls are but slightly eroded by the discharge, but when using large discharges, the surfaces are heated and worn away rapidly. For this reason, in the case of dischargers for large power it is necessary to rotate the balls. The author has devised arrangements in which two large metal balls or discs are enclosed in a chamber. Each of these surfaces is rotated slowly by means of an electric motor, and the discharge takes place between these discs or balls. The chamber can be filled with nitrogen or carbon dioxide gas, either at normal or under increased pressure. The erosion is then made uniform, and the discharge surfaces are eaten away only at one place. When it is desired to obtain high voltage sparks which are oscillatory, it is better to split up the spark between a number of balls in series, and not to unduly lengthen a single spark.

A matter of some importance in connection with the electric sparks used in radiotelegraphy is the resistance of the spark itself. It is a form of conductor, and as such possesses energy dissipating power or resistance, which is determined by its length and by the quantity of electricity it conveys.

The discussion of the various methods for measuring spark resistance must be deferred to the chapter on Radiotelegraphic Measurements. Meanwhile it may be said that whilst some physicists have measured the resistance of single sparks or discharges, these measurements do not possess much interest from a radiotelegraphic point of view. The matter of practical importance is the contribution which is made towards the total damping effect of a circuit containing a spark gap by the resistance of a rapidly repeated oscillatory spark. In the case of a damped oscillation the spark resistance is not constant during the successive oscillations, depending as it does upon the quantity of electricity carried across the gap. The resistance of an electric spark has this peculiarity, that the greater the quantity of electricity conveyed across in the discharge, the less is the true resistance of the spark. This is probably connected with the phenomenon characteristic of the electric arc and vaporous conductors generally, that the potential difference of the ends of the conductor decreases with increase in the current flowing through it. Hence, the ratio of P.D. to current decreases as the current increases, that is, the resistance decreases as the current increases. In the case of rapid trains of oscillatory sparks, the important quantity is the mean spark resistance during the continuance of the discharges. We may vary the spark length and yet keep the current constant, or we may keep the spark length constant and vary the current. If the current or discharge is kept constant, then the spark resistance appears to increase with the spark length and decrease with the capacity C in the manner shown by the curves in Fig. 16, which embody the results of some experiments by Prof. A. Slaby, taken by a method devised by K. Simons, in 1904, in which a metallic resistance is substituted for a spark of a given length and adjusted until the current in the circuit is the same as before. Similar experiments were made by the author, who

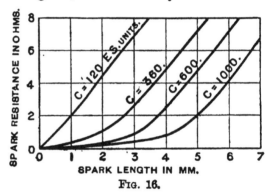

FIG. 16.

found that the spark resistance is to some extent determined by the material of which the spark balls are made. Measurements of spark resistance by another method called the resonance method, have been made by Rempp and Drude. Rempp gave curves which appeared to show that the spark resistance decreased at first with increasing spark length, and reached a minimum for sparks of about 6 mm. in length, after which it rapidly increased. Rempp's results have, however, been criticised by by Eickhoff, who has shown that Rempp neglected to take account of the effect of increasing spark length in causing brush discharges from the edge of the tinfoil of the Leyden jars used as condensers. This brushing involves loss of energy, and, unless a correction is made for it, it is reckoned in with the true dissipation of energy due to the spark, and thus credited as due to spark resistance.

Eickhoff has shown that if Rempp's results are suitably corrected
they show that with increasing spark length the spark resistance,
in an oscillatory discharge circuit comprising a condenser and
inductance, decreases at first rapidly and then tends to become
constant.   These results have been confirmed by the author, for

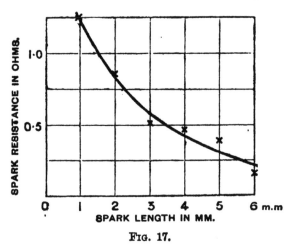

Fig. 17.

an oscillatory circuit comprising a capacity of 0·00261 mfd., an
inductance of 0·00636 millihenry and a spark gap varying from
1 to 6 millimetres, as shown by the curve given in Fig. 17.   The
larger the capacity used, the
lower is found to be the resistance of a spark for given
length.   The resistance of
the spark is therefore not
merely a function of its
length but is dependent upon
the circumstances under
which it is formed, and is
less the larger the quantity
of electricity which is carried
by it.

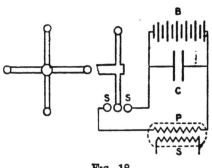

Fig. 18.

We have already referred
to the necessity for preventing an arc discharge accompanying the
true spark discharge of the condenser in those cases in which a
rapid series of oscillatory sparks is required, and especially when
using transformers and large condensers.   This can be achieved to
some extent by blowing air upon the spark gap, but even better
by employing as the discharger surfaces in very rapid relative

motion. This causes them to keep cool and completely prevents the arc discharge, whilst still permitting the condenser oscillatory discharge.

The arrangement in Fig. 18, due to R. Grisson, shows one mode of causing the oscillatory discharge of a condenser charged by a dynamo or battery to take place without arcing. In this arrangement the discharge is caused to take place between the balls SS by the periodic passage between them of four other balls carried on the arms of a rapidly rotating cross. Mr. Marconi has also devised several forms of non-arcing discharger to which we shall refer in connection with descriptions of radiotelegraphic apparatus and stations.

6. **Condensers for Electric Oscillations.**—The next element in the apparatus for the production of electric oscillations to be considered is the *condenser*. It consists of a sheet of a dielectric or insulator the two surfaces of which are nearly covered with some sheets of metal. The dielectric, which may be air, glass, ebonite, oil, or mica, is the real seat of the energy storage. The nature of the metal coatings is immaterial. They simply serve to discharge the dielectric when connected together. The simplest, and for some purposes the best form of condenser is an air condenser made by placing two sheets of flat metal parallel to each other and at a small distance apart. If this distance is d centimetres, and if the area of each plate is A square centimetres, then provided d is small compared with $\sqrt{A}$ we can say that the capacity of the condenser is very nearly $\dfrac{A}{36 \times 10^5 \times \pi d}$ microfarads.

An air condenser possesses no sensible conductivity or source of internal energy waste, but if the plates are near together and the potential difference large, a brush discharge will take place between the edges of the opposed plates which implies some dissipation of energy. Nevertheless, air condensers are very efficient oscillatory condensers. The chief objection to them is their bulk. The dielectric strength of air is about 38,000 or 39,000 volts per centimetre. Hence, if the condenser plates are 1 cm. apart, the above voltage cannot be exceeded without discharge. The energy put into a condenser of capacity C mfds. when charged to V volts is $\dfrac{1}{2}\dfrac{CV^2}{10^6}$ joules. Hence for such a parallel plate air condenser, we cannot put into it more than $\dfrac{A \times (38)^2}{36 \times 10^5 \times 2\pi d}$ joules, where A is the area of the surface of one plate and d their distance. If, then, the plates are 1 cm. apart, each must have

F

a surface 15,664 square cms. to enable the condenser to store up 1 joule or about 0·74 of a foot-pound of energy. Accordingly, two square metal plates, each 125 cms. in the side, placed 1 cm. apart, will store up 1 joule of electric energy as a maximum. At this rate we cannot store more than 1 foot-pound of energy in the form of electrostatic strain in 1 cubic foot of air. If, however, we employ compressed air as the dielectric of a condenser, as first suggested by R. A. Fessenden, we can then increase the dielectric strength almost proportionately to the pressure, and so store up in the same bulk of condenser much greater electric energy.

For most purposes, however, it is better to employ a solid or liquid dielectric, such as glass, ebonite, or some insulating oil. These all have a higher dielectric constant and a higher dielectric strength than air. Thus, for instance, good crown glass has a dielectric constant not far from 7, and for thicknesses such as 0·1 inch a dielectric strength of 200,000 volts per centimetre. Accordingly, it has five times the dielectric strength, and therefore $7 \times 5 \times 5 = 175$ times the energy storage capacity of air at atmospheric pressure for the same volume. Oil and ebonite have dielectric constants of about twice that of air, but ebonite has a greater dielectric strength than glass, and oil about half its value. Accordingly, a condenser for use in oscillatory circuits may be made as follows : Plates of thin zinc or brass are cut with a tail, as in Fig. 20, and these are arranged with tails on opposite sides on either side of a sheet of glass, say 0·1 inch or 2 to 3 mms. in thickness. Alternate sheets of glass and metal plates are built up, arranged as shown in Fig. 19. The glass is cut rather larger

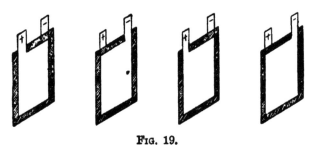

Fig. 19.

than the metal sheet, so as to leave a margin all round the metal plate. This pile of glass and metal sheets is enclosed in a stoneware or wooden box and covered in with vaseline or resin oil. In place of glass, sheets of ebonite may be employed. The immersion in oil is necessary to prevent sparking round the edges of the plates. If the glass is 0·1 inch thick and the metal sheets are

12 inches square, or have an area of 1 square foot, the single plate of glass included between two metal plates will have a capacity of $\frac{1}{500}$ of a microfarad.

A common form of condenser is the Leyden jar, consisting of a glass bottle or cylinder of thin English flint glass, lined inside and outside with tinfoil for about two-thirds of its height. The exposed bare glass surface must be well coated with shellac varnish, and the jar is closed by a wooden disc through which passes a brass rod terminated on the outside by a ball or terminal, and on the inside by expanding brass wires, which make a good contact with the inner tinfoil. Leyden jars for radiotelegraphic purposes should be carefully tested for dielectric strength, and also have their capacity in fractions of a microfarad marked on them. Leyden jars are combined into Leyden batteries by placing them in a box and joining them up either in parallel or in series. In the first case all the outside surfaces are connected metallically together and also all the insides ; and the capacity of the whole lot is then the sum of the capacities of each singly. In the second case the outside of one jar is connected to the inside of the next, and so on for a number of jars. This last mode of connection is called *cascade* connection, and then the reciprocal of the whole resultant capacity is equal to the sum of the reciprocals of each jar taken separately. Hence, if C_1, C_2, C_3, represent the capacities of three jars or condensers, and C the resultant capacity, when joined in parallel we have $C = C_1 + C_2 + C_3$, but when joined in cascade or series we have—

$$\frac{1}{C} = \frac{1}{C_1} + \frac{1}{C_2} + \frac{1}{C_3}$$

$$\text{or } C = \frac{1}{C_1^{-1} + C_2^{-1} + C_3^{-1}}$$

The reader will therefore notice that the rule for adding condensers in parallel is the same as that for adding resistances in series, and for adding capacities in series is the same as that for adding resistances in parallel.

Moscicki has devised a kind of Leyden jar consisting of a glass tube, of which the walls are rather thicker at the ends than in the middle, being gradually increased in thickness in passing from the centre to the ends. These tubes are coated inside and outside with a silver deposit electroplated over with copper for strength, and the coatings extend up to a certain distance from the ends. These tubes are combined in batteries and form convenient high-tension condensers. The reason for thickening the glass at the ends is because a Leyden jar is apt to pierce when charged too highly at or near the edges of the coatings, at which place the electric force is greatest. Moscicki found that a coated plate of glass 0·5 mm. in thickness would stand 67,000 volts without rupture in its centre portions, but that it was pierced by 11,000 volts near the edges of the tinfoil coatings.

If a Leyden jar is used as the condenser in an oscillatory circuit, then when the charging potential is high enough a brush discharge or glow is seen along the edges of the tinfoil, and this implies a discharge into the air. It causes a small increase in the effective capacity of the jar, which is larger under high charging potential than under low. It is prevented by immersing the jar in insulating oil. A standard condenser is best made by employing metal plates placed a few millimetres apart in high flash-point paraffin oil free from every trace of water as the dielectric.

A number of flat zinc or brass plates are prepared of the same size and are separated to a small distance by inserting the edges in saw cuts made in slips of ebonite. Alternate plates are connected by wires. The whole mass of plates is suspended in a glass vessel filled in with the paraffin oil, and the capacity is determined as described in Chapter VIII. of this book. Such a condenser possesses no sensible dielectric energy loss, and the oil doubles the capacity of the condenser as compared with its capacity when in air, because the dielectric constant of the oil is nearly 2.

7. Inductances and Oscillation Transformers.—The third element in the oscillation circuit is the inductance coil or oscillation transformer. To produce the high frequency oscillations required in radiotelegraphy, we must employ a circuit having inductance in series with capacity. One form which the inductance may take is that of a loop of one or a few turns of insulated wire. In one shape it may be constructed by making a square wooden frame 12 inches, or 30 cms., or more in side, and 4 inches, or 10 cms., in width.

On this is wound a number of single turns of cable joined in parallel, each turn being a rope made up by twisting together a large number of fine copper wires, each insulated with cotton. The object is to obtain a coil of one turn which shall have the least possible effective resistance, and also small inductance. As we have seen in Chapter I., to reduce the resistance we must prevent the skin effect and compel the current to make use of the interior as well as of the surface of the wire, and to reduce the inductance we must employ only one turn of wire.

We may wind over this single primary turn, which is placed in series with the condenser and the spark gap, another winding of several turns also made of insulated wire, and thus construct what is called an *oscillation transformer.* When the primary coil is traversed by oscillations they induce other oscillations in the secondary circuit, but these last are feeble unless a condenser is put across the secondary terminals and the two circuits brought into resonance. If, however, this is done, then the primary oscillations induce secondary oscillations, and by a suitable choice of number of turns and capacity we may obtain from across the terminals of the secondary condenser a much higher potential difference than that between the terminals of the primary condenser. These two syntonic circuits, moreover, as we have shown at the end of the previous chapter, act and react on each other in a curious manner, so that the result of coupling together two circuits inductively (called by the German writers magnetic coupling) is to create oscillations of two periods in both circuits, one greater and the other less than the free natural period of either circuit taken separately.

We have given already a mechanical illustration by an experiment with two pendulums of the reasons for this production of oscillations of two periods, but it is desirable to elucidate the matter more completely. We shall assume that the capacity in the primary oscillation circuit is C_1 microfarads, and that the inductance is L_1 microhenrys, and that the capacity and inductance in the secondary circuit are C_2 and L_2 respectively. The two circuits are assumed to be syntonised so that $C_1 \times L_1 = C_2 \times L_2 = CL$, and the natural time period of oscillation of either circuit taken alone is $T = 2\pi\sqrt{CL}$ microseconds.

Let V_1 and V_2 be the maximum values of the potential differences of the terminals of the primary and secondary condenser respectively, and let I_1 and I_2 be the maximum values of the currents in the two circuits respectively. Then, if we consider the resistances of the circuits to be negligible, the current (max. value) at any instant in one circuit is equal to the rate at which the charge of its condenser is varying. The charges of the two condensers are the products C_1V_1 and C_2V_2 respectively. We have shown in section 6 of Chapter I. that if $p = 2\pi n$, where n stands for frequency, that the maximum value of the time rate of change of any simple periodic quantity is p times its maximum value. Hence the currents (max. values) in the two circuits are given by the equations $I_1 = pC_1V_1$ and $I_2 = pC_2V_2$ respectively. Also, if L_1 and L_2 are the inductances of the circuits, the magnetic flux linked with either circuit due to its own current is measured by L_1I_1 and L_2I_2 for the two circuits respectively, and therefore the counter electromotive forces due to inductance, or, as it is called, the reactance voltages, in each circuit are pL_1I_1 and pL_2I_2 respectively. Again, each circuit not only is linked with its own lines of flux, but with some of those of the other. Let M be a quantity, called the *coefficient of mutual inductance*, such that MI_2 is the flux due to the secondary circuit which is linked with the primary, and MI_1 is the flux due to the primary which is linked with the secondary circuit. Then pMI_2 and pMI_1 are the induced electromotive forces in the primary and secondary circuits respectively.

Now, in each circuit the potential difference of the condenser terminals is the sum of the reactance voltage and induced voltage. Hence we have for the two circuits the electromotive force equations

$$V_1 = pL_1I_1 + pMI_2$$
$$V_2 = pL_2I_2 + pMI_1$$

Eliminating I_1 and I_2 by the help of the equations $I_1 = pC_1V_1$, $I_2 = pC_2V_2$, we have finally the equations

$$(1 - p^2L_1C_1)V_1 + p^2MC_2V_2 = 0$$
$$p^2MC_1V_1 + (1 - p^2L_2C_2)V_2 = 0$$

Furthermore, eliminating V_1 and V_2 by cross multiplication, we arrive at the biquadratic equation

$$C_1C_2(L_1L_2 - M^2)p^4 - (C_1L_1 + C_2L_2)p^2 + 1 = 0$$

If we write k^2 for $\dfrac{M^2}{L_1L_2}$, and if we assume the two circuits are tuned so that $C_1L_1 = C_2L_2 = CL$, then, making these substitutions in the above equation, it is easy to see that

$$p^2 = \frac{1}{CL}\frac{1 \pm k}{1 - k^2}$$

Hence there are two values of p, according as we take the positive or negative sign. Since the natural frequency n of either circuit alone is given by the equation $n = \dfrac{1}{2\pi\sqrt{CL}}$, we may denote the two roots of the above equation by the symbols $p_1 = 2\pi n_1$ and $p_2 = 2\pi n_2$, and then we have

$$n_1 = n\frac{1}{\sqrt{1 - k}}$$

$$n_2 = n\frac{1}{\sqrt{1 + k}}$$

The quantity k is called the *coefficient of coupling*, and it denotes the ratio of the coefficient of mutual inductance M to the square root of the product of the separate inductances L_1 and L_2 of the two circuits. M is always numerically less than $\sqrt{L_1 I^2}$ because the number of lines of flux a circuit can send through a neighbouring circuit is less than those linked with itself. Hence k is a proper fraction, and $\sqrt{1-k}$ is less than unity, and $\sqrt{1+k}$ is greater. Hence we see that n_1 must be greater than n, and n_2 less than n. Accordingly we derive the following important conclusion. When two oscillatory circuits are inductively coupled together so that oscillations in one excite oscillations in the other, and if these circuits are tuned to the same frequency when separate, then when coupled together, oscillations of two frequencies are set up in them both, that of one being greater, and that of the other less, than the natural frequency of either when alone.

The difference of these two frequencies depends upon the coefficient of coupling of the circuits. If k is a small fraction, the circuits are said to be *loosely coupled*, and then n_1 and n_2 are not very different. If k is a large fraction near to unity, then n_1 and n_2 are very different, and the circuits are said to be *closely coupled*.

For many purposes a very convenient form of variable inductance may be made as follows: two cylinders of hard dry wood or of ebonite have a coarse screw groove cut on their surface with a pitch of 8 or 10 to the inch. In this groove, bare copper wire, No. 14 or No. 16 S.W.G., is wound, and the ends of the wire attached to screw terminals in the ends of the cylinders. The groove should be so deep that the wire lies halfway deep in it. These cylinders are mounted up on a board side by side. A thick and heavy strip of copper is bent as in Fig. 20, so as to fit the

Fig. 20.

curved surface of both parallel cylinders and lie on them. An ebonite handle serves to slide it along. The current enters at one end of one spiral, passes down it a certain distance, then cuts across the copper strap to the other, and returns by the other spiral. Hence, by moving the strap more or less along the cylinders, a variable amount of inductance can be inserted in any oscillatory circuit. The inductance of any length of the spiral can be predetermined by the formula given in Chapter I., in the section on Inductances.

Another mode of making an inductance which can be varied, is to construct two circular coils of equal number of turns, one of which is rather smaller than the other and can revolve on an axis arranged as a diameter to the larger coil. The two coils are joined in series with each other, and when they are arranged in the same plane with the windings following on in the same direction, they have their maximum inductance. If, however, the inner coil is turned round so as to be at right angles to the outer, or turned right about face, so that the windings are in the opposite direction, then the two coils in series have a greatly reduced joint inductance. By the use of appropriate induction coils, we can transform electrical oscillations, increasing the current and lowering the potential, or *vice versâ*.

Oscillation transformers for creating extra high potential discharges by means of lower potential oscillations are often called Tesla coils, though as a matter of fact employed by many physicists prior to the date of Tesla's researches. A coil of this description consists of a primary circuit, which should have few turns, and a secondary circuit of many turns. These circuits must be constructed of highly insulated wire, and the secondary circuit should be preferably formed of one single layer of wire wound on an ebonite or glass tube. The primary circuit is in series with a condenser and spark gap. The whole coil must be immersed under some highly insulating oil to prevent brush discharges. When oscillatory discharges are sent through the primary circuit, oscillations are created in the secondary circuit of higher potential, and long sparks and powerful electric brush discharges can be taken from the ends of the secondary circuit. The effect may be increased by " tuning " the two circuits by adding capacity to the secondary circuit.

8. **Multiple Transformation of Oscillations for High-Tension Condenser Charging.**—We can employ the high frequency oscillations produced by a Tesla coil or oscillations transformer to charge an oscillation circuit, and so produce trains of oscillations which are not only of great amplitude, but succeed each other with great rapidity.

The following arrangement, devised by the author in 1900, is such a system of multiple transformation for high potential condenser charging. A high-tension alternator, D (see Fig. 21), provides an alternating current having a frequency, say, of 50 at a pressure of 2000 volts. This current passes through the thick wire of an ordinary high-tension transformer, T^1, and is transformed up to 20,000 or 30,000 volts. Across the secondary terminals of this transformer are connected a pair of spark balls, S¹, a condenser, C¹, and the primary coil of an oscillation transformer, T². The secondary circuit of this last is connected in turn to a pair of spark balls, S², and to a condenser, C², and the primary circuit

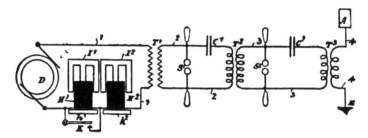

Fig. 21.

of a second oscillation transformer, T³. The secondary circuit of this last transformer then provides oscillatory discharges of extra high tension and high frequency, and a large number of trains of oscillations per second.

The operation of the apparatus is as follows: At each alternation of the current in the alternator, a current traverses the first transformer T¹, and creates alternations of potential which charge the condensor C¹. If the circuit composed of the secondary circuit of the transformer T¹, the primary circuit of the transformer T² and the condenser C¹ has its capacity and inductance adjusted to be in resonance with the low frequency (say 50) of the alternator, then powerful oscillations will accumulate in it, which at intervals will discharge across the spark gap S¹. Thus there will not be 100 sparks per second at S¹, corresponding to the 50 frequency, but perhaps 10 or 12. At each of these sparks the condenser C¹ discharges with oscillations and gives rise to a long train of damped oscillations. These are transformed up in potential by the transformer T², and in like manner charge the condenser C², and, if the circuit of this condenser is properly tuned to the circuit of the condenser C¹, then, in like manner, powerful oscillations are set up in the circuit composed of C² and T³, and when sparks occur at the second spark gap S₂ we have high potential high frequency oscillations in the circuit of C² which consist of multiple trains of oscillations, a group of trains in the circuit of C² corresponding to each one in the circuit of C¹. Special means have to be provided for preventing the arcing at the primary spark balls, which will be described in a later chapter.

We have already shown that when two oscillatory circuits are in tune and coupled together inductively, oscillations of two frequencies are created in them by their mutual reaction. Hence in the above-described arrangement the effect produced in the last oscillation circuit is a very complex one, and cannot be described as a simple series of trains of oscillations of one period. Other arrangements with multiple spark gaps have been devised, by which oscillations can be created in definite relative phase difference to each other, such, for instance, as that due to Mandelstam and Papalexi, which is as follows:

A circuit is constructed as in Fig. 22, which contains two con-

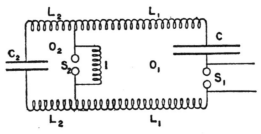

FIG. 22.

densers, C_1 and C_2, two spark gaps, S_1 and S_2, and so arranged as to form two oscillation circuits, O_1 and O_2. The spark gap S_2 is short-circuited by a large inductance, L, and the other portion of the oscillatory circuits comprise inductances L_1 and L_2. If, then, the spark gap S_1 is connected to an induction coil or transformer, the inductance L offers no obstacle to the slow charging of the condenser C_1, which accordingly becomes charged, but the condenser C_2 is not charged. When the potential of C_1 reaches a certain value, it discharges across the spark gap S_1 with oscillations which take place in the circuit comprising the two condensers and the two inductances with a frequency determined by the inductances L_1 and L_2 and the capacities C_1 and C_2. At a certain phase of the discharge determined by the constants of the circuit a discharge takes place across the spark gap S_2, and sets up independent oscillations in the circuit O_2, which have a frequency determined only by the capacity and inductance C_2, L_2. Hence we have oscillations started in the circuit O_2 which bear a definite relation to those in O_1 as regards phase of maximum value, determined by the inductances and capacities and the spark-gap length S_2. The full theory of the action can only be explained by the aid of mathematical analysis, but a comprehension of the principles involved may be obtained by considering a mechanical analogy.

Imagine a heavy weight suspended by means of a spiral spring from a fixed support. If the weight is pulled down and released, it vibrates up and down at a rate determined by the mass of the bob and the stiffness or extensibility of the spring. The mass of the bob corresponds to the inductance of an electrical circuit and the extensibility of the spring per unit force to the capacity of the condenser and the mechanical rate of vibration to the electrical frequency. If we suppose the weight pulled down by a thread which breaks when the tension in it exceeds a certain value fixed by the extension of the spring, we may regard this thread as acting like the air in the spark gap in suddenly releasing the strain and starting the oscillations.

Next suppose that a second weight of different mass is hung alongside of the first, suspended by a spring of different extensibility, and let the two weights be connected through a pivotted lever, as in Fig. 23. The left-hand weight and spring must be supposed to be connected to the lever by a thread which will be ruptured by a certain tension. Hence on pulling down the right-hand mass by the thread S_1 the corresponding spring is extended and the first system alone stores up strain. When the thread S_1, in Fig. 23, corresponding to the spark gap S_1, in Fig. 22, snaps, the right-hand mass L_1 oscillates, and it communicates its oscillations to the left-hand mass L_2, and as these systems are not in tune there is a strain brought to bear on the thread S_2, which finally snaps and releases the mass L_2, which thereafter executes free oscillations in a period determined by the mass L_2 and the resilience of the spring C_2. In this manner the free oscillations

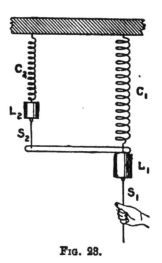

Fig. 23.

of the system L_2C_2 are started by the oscillations of the system L_1C_1.

We shall refer in Chapter V. to a system of directive telegraphy by Prof. F. Braun in which this method has been applied for the production of two trains of oscillations having fixed phase relations. The special claim made by A. Jollas for this method of charging devised by Mandelstam and Papalexi is that by this means it is possible to convey to a condenser a greater energy than corresponds with the length of the spark gap used. In the ordinary or single oscillation circuit we cannot give to the condenser more energy than $\frac{1}{2}CV$ joules, where C is its capacity in farads and V is the spark potential corresponding to the length of spark gap l employed. Hence the attempt to increase V involves an increase in l, and this involves an increase in the resistance of the spark, and therefore a larger damping in the circuit, and therefore fewer oscillations per train. Hence the *integral effect* of the oscillations as estimated by the heating effect of the whole of the train when passed through a fine wire is not necessarily increased, but may be decreased by increasing the spark length. We diminish the number of oscillations in a train if we try to increase the amplitude of the initial oscillation. It appears therefore that there is a certain length of spark gap which gives the least damping and therefore greatest integral effect, and this appears to correspond to a very short spark length in air of about 1 mm. or less.

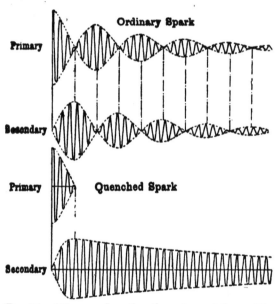

Fig. 24.—Diagram denoting the nature of the oscillations in the primary and secondary circuits for an ordinary and for a quenched spark.

9. Quenched Spark Methods for the production of Damped Oscillations.—Recent methods for the production of damped oscillations generally employ some form of quenched spark in which the spark resulting from the discharge of a condenser is rapidly extinguished. This method is based upon a discovery made by M. Wien in 1906 on the damping of short sparks. If a condenser is discharged across a spark gap which is more than a millimetre or so in length, then provided there is sufficient inductance and capacity in the circuit and also not much resistance, the discharge is prolonged and a series of oscillations takes place before the condenser is finally discharged. If then this primary circuit is coupled inductively to a secondary circuit the two act and react on each other as already described, with the result that the oscillation in each circuit is of a complex nature and can be resolved into two sets of oscillations of different frequencies. If, however, the spark gap is very short and consists of two smooth plates of good conducting metal placed with their surfaces parallel to each other and less than a millimetre apart, then the condenser spark is rapidly extinguished or quenched, and as far as the condenser circuit itself is concerned the discharge becomes nearly dead beat or at most consists of one or two oscillations. If then this primary circuit is coupled to a secondary circuit, the quenched discharge in the former administers to the latter an electrical impulse which results in the production in the secondary circuit of its own free period oscillations or of trains of many oscillations of one period only.

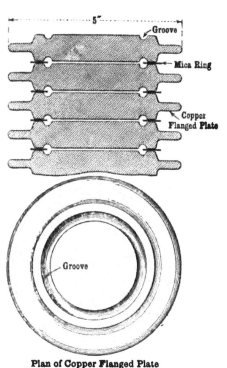

The nature of the oscillations in the two circuits in the case of the ordinary and of the quenched spark are illustrated by the wavy lines in Fig. 24.

There are many ways in which this quenched primary spark can be produced. One form of quenched spark discharger consists of a series of thick copper plates the surfaces of which are made very smooth and parallel. In each plate a deep groove is turned, and a mica ring is laid between every pair of plates so as to introduce a gap between them. The discharge then takes place across this narrow gap about 0·25 mm. in width. The object of the groove in the plates is to prevent the discharge taking place at the edge of the mica ring (see Fig. 25). A pile of such plates can be arranged so that the

Plan of Copper Flanged Plate

FIG. 25.—Construction of a quenched spark discharger.

gaps are in series resulting in the use of a large total voltage to form the sparks (see Fig. 26). This form of quenched spark discharger is called a Wien discharger. The surfaces must be very smooth, parallel, and near together in order that the spark may be quenched by the cooling action of the metal surfaces. A very similar form of quenched spark discharger was invented by Von Lepel in which a paper ring is used to separate slightly two water-cooled flat or conical metal surfaces between which the discharge takes place.

Fig. 26.—A quenched spark discharger made as shown in section in Fig. 25.

Another form of quenched spark discharger is that due to Peukert, in which a thin film of oil between two rotating metal surfaces is pierced by the spark. In the Wien form of discharger with fixed plates, there is a tendency for the spark to continue to pass at the same points between the surfaces, and thus to eat them away at certain points and increase the distance between them. This is overcome by making the surfaces rotate.

A modification of Peukert's discharger invented by the Author, is a very effective form of quenched spark discharger, and of great use in laboratory work. It is constructed as follows :—

Two very flat plates, A_1, A_2 (see Fig. 27), of hardened steel are prepared, and one of these has a small hole in the centre.

These plates are held in a frame which can be immersed in a vessel of paraffin oil. One of the plates is attached to a steel shaft which runs in ball bearings, and can be rotated by an electric motor. The plate is fixed with its plane perpendicular to the axis of rotation. The other plate is held by adjusting screws with its surface parallel and very near to that of the rotating plate (see

Fig. 27). Hence there is a thin film of oil between the plates. When the upper plate revolves, this oil is flung out, but continually renewed by suction through the central hole in the bottom plate. The two plates are electrically insulated, and form the surfaces between which the discharge takes place. A number of these pairs of plates can be placed in series. This discharger is very efficient as a quenched spark discharger, and as the spark continually changes its place, the plates are worn away equally and not pitted in one place.

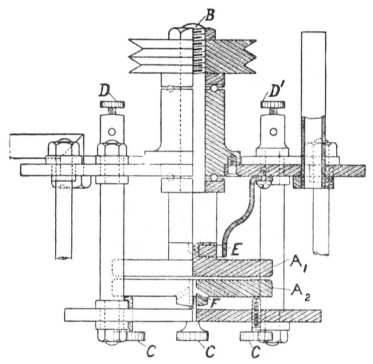

Fig. 27.—Section of Fleming quenched spark discharger.

The above form of discharger when inserted in a condenser circuit, produces a very rapid series of strongly damped discharges, and these create a similar series of feebly damped discharges or trains of oscillations in a coupled secondary circuit, these latter having the frequency natural to the secondary circuit.

On the other hand, there is nothing to control the exact spark or train frequency.

Dischargers in which there is an extremely regular spark

frequency are called *musical spark* dischargers. These are generally constructed with one fixed electrode, and the other a revolving one with projecting pins, balls, or studs, which make grazing contact with the fixed electrode during their revolution.

Such, for instance, is the high speed musical spark discharger of Marconi, which is described in Chapter VII.

These musical spark dischargers have an especial advantage in the case of wireless telegraphy when using a telephone as a receiver, to which reference will be made in a later chapter.

CHAPTER III

UNDAMPED ELECTRIC OSCILLATIONS

1. **High Frequency Alternators.**—It has already been explained that undamped or persistent electric oscillations are extra high frequency alternating currents, the frequency of which may be from 1000 to 10,000 times greater than those of the so-called low frequency alternating currents used for electric lighting and the transmission of power. The production of these undamped oscillations has attracted great attention of late years, and several methods have been found by which they can be produced. One of these is by the use of a high frequency alternator. The invention of these machines dates back to about 1889 or 1890, when arc lighting by alternating currents began to attract attention, and it was hoped that by the employment of alternating currents of a frequency of 10,000 or more, the sound of the alternating arc which was very noticeable at 50 or 100 ~ would be annulled. Elihu Thomson and Nikola Tesla were successful in constructing such machines.

Tesla constructed one form of high frequency alternator as follows (see Fig. 1): It consists of a fixed ring-shaped field magnet with magnetic poles projecting inwards and a rotating armature in the form of a fly-wheel. This wheel, J (see Fig. 2), was turned down on the edge, forming a kind of flanged pulley, and this groove is wound full of annealed iron wire insulated with shellac. Pins, L, were set in the sides of the ring J, and flat coils, M, of insulated wire wound over the periphery of the armature wheel and around the pins. These coils were connected together in series, and the ends of the series carried through a hollow shaft, H, to slip rings, P, P, from which the currents were taken off by brushes, O, O. The field magnet consisted of a kind of toothed wheel, with the teeth turned inwards (see Fig. 2), and an insulated wire or strip was wound zigzag fashion between these teeth, so that when a continuous current was passed along this conductor, the teeth were

G

made alternately North and South magnetic poles. It is quite
possible thus to produce a magnet having 400 radial poles in the

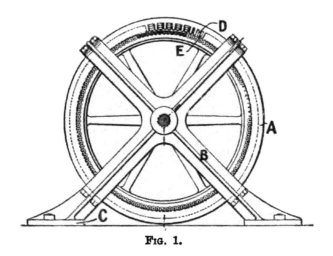

FIG. 1.

circumference and also easy to put 400 coils on the armature.
Hence if such a machine is driven at a speed of 3000 revolutions

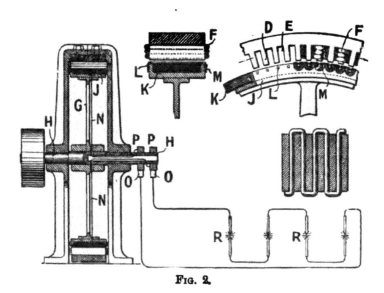

FIG. 2.

per minute, or 50 per second, it produces an alternating current
having a periodicity of 10,000 ~.   A machine of this kind can be

constructed to give a current of, say, 10 amperes. In the machine above described, which was capable of giving an alternating electromotive force of about 100 volts, the field magnet consisted of a ring of wrought iron 32 inches outside diameter, about 1 inch thick, the inside diameter was about 30 inches. The distance between the teeth was about $\frac{3}{16}$ inch, and each field magnet tooth was about $\frac{3}{16}$ inch thick. On the armature 384 coils were connected in two series. The width of the armature was $1\frac{1}{4}$ inch. With magnetic teeth placed so close it was necessary to have an extremely small clearance between the armature coils and the magnet, to avoid excessive leakage or loss of useful magnetic flux, hence, it was impossible to use wire for the armature thicker than No. 26, Brown and Sharp gauge. This size is equivalent to No. $28\frac{1}{2}$ British S.W.G. The armature wires must be wound with great care, otherwise they are apt to fly off in consequence of the great peripheral speed. It is practicable to run such an armature at a speed of 3000 revolutions per minute, equivalent to a peripheral speed of 375 feet per second.

In another type of machine constructed by Tesla, magnetic leakage was avoided by making adjacent poles on the same side of the armature of the same polarity. In this second form the armature consisted of a copper plate in the form of a disc with a large hole in it (see Figs. 3 and 4). The plate was cut through by radial slits alternately at the inside and outside edge, so as to divide the plate up into a zigzag strip. This plate was clamped on a central boss fixed on a shaft (see Fig. 4) and caused to revolve between the two parts of a field magnet having a large number of

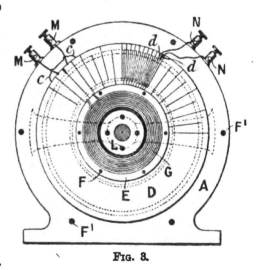

FIG. 8.

inwardly projecting poles, all those on one side being of the same polarity and facing an equal number of like poles on the opposite side, of the opposite polarity. In this manner, the disc was perforated by the magnetic flux passing across from one set of poles to another, and the passage of the strips into which the disc is cut up, into and out of these streams of magnetic flux gives rise to the electromotive force in the armature. The armature winding therefore consisted of a single disc-shaped conductor equivalent to a zigzag winding, and this was driven at a high speed so that the radial elements of the armature cut across streams of magnetic flux. A very strong excitation could therefore be employed without producing any wasteful leakage flux. The chief defect of this design of armature is that unless the slits in the disc are very close together, so that the width of the radial bar or slice is not more than $\frac{1}{32}$ inch, there is considerable heating of the armature, due to eddy currents set up in it. In one

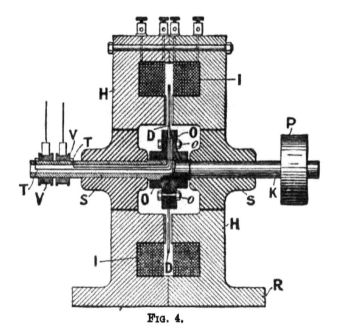

Fig. 4.

machine of this type, constructed by Tesla, the field had 480 polar projections on each side, and from this machine it was possible to obtain a current having a frequency of 15,000 complete periods per second. When a machine of this description having a disc of considerable diameter is driven at a speed of 3000 R.P.M., very accurate balancing is necessary, or otherwise dangerous vibrations will be set up in the machine. Great rigidity and accuracy of work is therefore necessary in all parts of the machine, because the clearance between armature and field magnets must necessarily be very small.

There is very great difficulty in securing the necessary balance in any armature having wound coils upon it.

Hence the inductor form of alternator has been adopted in most cases for high frequency machines. The revolving part is then merely an iron disc having teeth or notches cut on its edge. If two chisel-shaped magnetic poles are placed on either side of such a disc, and if these poles carry armature coils wound on them, then as the notched iron disc rotates it varies the magnetic reluctance of the magnetic circuit, and hence the flux passing through the armature coils. In this manner an electromotive force is created in them which has a frequency determined by the speed of the iron disc and the number of its teeth.

W. Duddell has described the construction of a high frequency alternator of the inductor type. It consists of a laminated soft iron ring having two inwardly projecting poles. This ring is wound with an exciting circuit, so that a direct current flowing in this circuit tends to make one of these poles North and the other South. In addition, another or armature circuit is laid upon the ring. Between the pole pieces a laminated soft iron disc revolves which has V-shaped notches cut on its periphery.

The exciting circuit on the ring had inductance coils inserted in it, so as to prevent high frequency currents being generated in it. The iron inductor disc was revolved by a cotton belt passing round a pulley on the inductor shaft and round two large metal disc pulleys which in turn were driven by an electric motor. In this manner the inductor disc was driven at 30,000 or 40,000 R.P.M. Alternating currents could be obtained from the armature circuit having a frequency up to 18,000 per second. The machine gave a current (R.M.S.) of 1 ampere and an electromotive force of 40 volts. Subsequently inductors with 50 or 60 teeth were used and driven at speeds up to 600 revolutions per second. This furnished an alternating current having a frequency of 50,000.

Finally an inductor disc was made with 204 teeth, merely a sort of laminated iron disc with a milled edge. Coils of wire were wound on the iron pole tips as armature coils, and with this arrangement it was finally found possible to create an alternating current having a frequency of 120,000 when the disc was driven at a speed of 600 revolutions per second. On the other hand, the output of the machine was then very small, being only 0·1 ampere at 2 volts. The alternator gave 3·6 volts on open circuit. This machine was constructed for experiments on the electric arc, and not primarily for the purpose of electric oscillation work.

Such a small output is, however, useless for the purposes of radiotelegraphy or telephony.

Inductor alternators, giving a considerable output, have, however, been constructed by E. F. W. Alexanderson of the General

Electric Company in the United States. The coils which serve as field coils and provide the stationary magnetic field are wound on pole pieces which form part of a fixed iron ring frame. There are certain other fixed coils which are the armature coils. Between these revolves a steel disc having teeth cut on its periphery which are filled in with brass so as to present a smooth surface. As this disc revolves the steel teeth serve to increase or diminish the amount of magnetic flux which passes through, and is linked with the armature coils. This creates in the latter an electromotive force. Hence alternating currents are generated in these armature coils although the only moving object is the steel disc. Alexanderson has in this manner constructed alternators giving an output of 2 kilowatts at a frequency of 100,000, and of 35 kilowatts at a frequency of 50,000. This is quite high enough for radiotelegraphy.

A new method of generating high frequency currents was, however, described in 1907 by Dr. R. Goldschmidt. His machine enables us to increase frequency so that starting with an alternating current having a frequency, say, of 10,000, we can create by it another current having a higher frequency of 40,000 or 50,000. The point of practical importance in this appliance is that it does not involve extraordinarily high speeds of rotation in the rotating part even when currents having a frequency of 40,000 to 100,000 are being generated.

The principle on which this invention operates may be explained as follows :—

If we consider in any space a steady uniform magnetic field, we may suppose these lines of magnetic force to turn round a centre like the hands of a watch without changing their strength or intensity. Such a rotating magnetic field is called a revolving field. On the other hand, if we have an alternating current of electricity flowing in a coil, it produces a magnetic field constant in direction, but varying in strength periodically. If the field strength varies so that the strength at any instant is to the maximum strength in the ratio of the sine of the angle $2\pi nt$ to unity, where n is the frequency or number of complete cycles per second, and t is the time reckoned from the instant of zero value of the field, then the field is said to vary harmonically, or to be a simple periodic field of constant direction. Now it is easily shown that such a field is the resultant of two equal steady revolving fields rotating in opposite directions each of half the maximum strength of the periodic field, the angular velocity of the rotating field being equal to $2\pi n$. Accordingly we may consider any unidirectional but periodic field to be resolved into two oppositely rotating but steady fields (see Fig. 5).

Suppose, then, that we have a closed coil of wire rotating in a steady magnetic field. We have produced in this coil an alternating electromotive force and current, and also an alternating or periodic magnetic field which has a constant direction, viz. along the axis of the coil. This periodic field can be resolved into two oppositely rotating constant fields, one of which rotates in the same direction as the coil, and one in the opposite direction. One of these rotating fields moves through space with twice the angular velocity of the coil, and the other is stationary as regards an external object, say the stationary electro-magnets producing the exciting field.

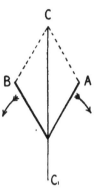

FIG. 5.

This stationary component acts against the steady exciting magnetic field, and provides what is called the armature reaction tending to weaken the impressed field. The other rotating component, revolving forwards with twice the angular velocity of the armature, cuts through the exciting coils of the field magnets and induces in them an alternating current of twice the frequency of the armature current. This current produces an alternating field which in the same manner can be resolved into two oppositely rotating fields of constant strength having twice the frequency of the armature or rotor current. Hence both of these cut through the rotating coils and produce currents having respectively the fundamental frequency and three times that frequency. Let us suppose, then, that both the field magnet coils or so-called stator coils, and also the rotating armature or rotor coils, are closed by circuits consisting of an inductance coil of very low resistance in series with a condenser (see Fig. 6). This inductance (L) and capacity (C) can be adjusted or tuned to a frequency n or an angular velocity $p = 2\pi n$ by the rule that $p = \dfrac{1}{\sqrt{LC}}$.

Let us suppose that the angular velocity of the armature or rotor coil is p, and that the inductive-capacity circuit completing it is tuned for a frequency $n = \dfrac{p}{2\pi}$. Also let the stator coil be short-circuited or completed by an inductive-capacity circuit tuned to a frequency $2n$, and let the rotor coil have also a second inductive-capacity circuit tuned to a frequency $3n$ (see Fig. 6).

We have then provided external circuits in which the

alternating currents can flow which are circulating in the rotor and stator coils.

The above description gives an outline of the operation of the Goldschmidt alternator for the production of high frequency currents. It will be seen that this machine multiplies the frequency of an alternating current. The actual alternator consists of a fixed portion called the stator, which consists of a number of coils fixed on laminated iron pole-pieces attached to a frame, and these coils are traversed by a continuous current which provides an exciting magnetic field (see Fig. 7). In the interior of this field-magnet frame a laminated slotted iron core revolves, in the slots of which is wound a zigzag winding connected at the ends to a pair of slip rings on the shaft, so that connection may be made with the condenser-inductive circuits outside. The rotor is made to revolve at a high speed, and generates in these

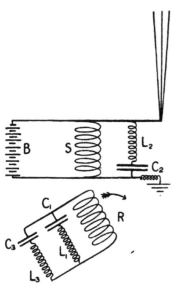

FIG. 6.—Diagram connections of Goldschmidt alternator.

coils an alternating current having a frequency, say, of 10,000. This in turn creates in the field or stator coils an alternating current having a frequency of 20,000 in virtue of the actions above explained. We have then an additional pair of currents created in the rotor coils having frequencies of 10,000 and 30,000 respectively, which circulate in the two tuned inductive-capacity circuits connected to the rotor.

Again the currents of frequency 10,000 and 30,000 act to produce in the stator currents of 20,000 and 40,000 respectively. If an external circuit tuned for 20,000 has been connected to the stator, the current of this frequency can be taken up in it. Lastly, if an aerial wire or antenna is connected to the stator tuned for a frequency of 40,000 by proper adjustment of its inductance, is connected to the stator, we should be able to set up in it continuous electric oscillations of this last frequency.

One important effect remains to be noticed. It will be seen that there are pairs of currents of the same frequency which occur in the rotor and stator. Thus for the frequency 10,000, there is one current induced in the rotor by its rotation in the steady or

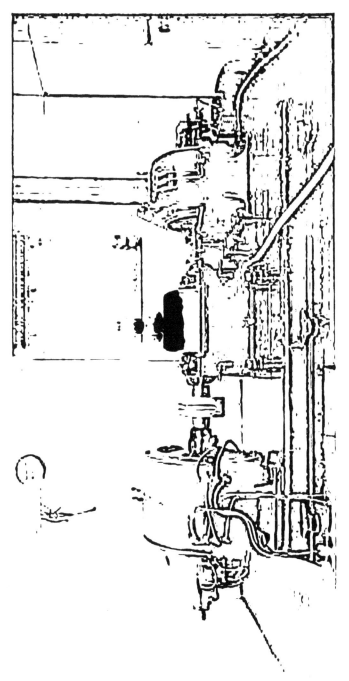

Fig. 7.—View of Goldschmidt's high frequency alternator.

exciting field of the stator, and one current of the same frequency produced by the reaction of the current of frequency 20,000 occurring the stator. Also for the frequency 20,000 there are a pair of currents in the stator. These pairs of currents are nearly in opposite phase, and hence more or less nullify each other. Accordingly the only unbalanced current is the final current of highest frequency.

Hence by means of an alternator in connection with capacity shunts, as they are called, Goldschmidt was able to provide the means of mechanically generating extra high frequency currents. As it is not very convenient to employ a large number of shunts, the number of steps by which the frequency is raised is in practice limited to 4 or 5. Therefore starting with an alternating current of 10,000, we can easily increase this to 40,000. A frequency of 40,000 corresponds to a wave length of 25,000 feet or about 5 miles. This is suitable for long-distance radiotelegraphy, but not for short or ship work. Hence Goldschmidt's machine, which can now be constructed of sizes up to 150 K.W. or more, is especially adapted for long-distance radiotelegraphy, but not for wireless telegraphy on a smaller scale. An external view of the Goldschmidt alternator is shown in Fig. 7.

There are methods of increasing frequency by means of static transformers which deserve a brief mention. One of these, due to Joly, enables us to obtain from a single-phase alternating current of given frequency another of triple frequency by means of two transformers.

Two iron-core transformers have their primary and secondary coils respectively joined in series. The primary turns of one transformer are so adjusted that the core never becomes magnetically saturated during the period, whilst those of the other are adjusted so that the core is saturated at an early stage during the period. If the secondary circuits are joined upon opposition and the turns on each transformer adjusted to create about equal E.M.F. in each secondary, the result will be to create a triple-frequency current in the secondary circuit, which can have its amplitude exalted by properly tuning that circuit.

Thus, if we supply two such transformers with a single-phase current having a frequency of 1000, we could by this means obtain a current having a frequency of 3000, and by a double transformation one having a frequency of 9000.

By the use of two transformers each having a third winding on its core transversed by a continuous current, we can create differences in the magnetic state of the two cores such that when the primaries and secondaries are joined in series the result of supplying the primaries with a single-phase current is to produce a double-frequency current in the secondaries. These devices have, however, not come into extensive use.

2. The Production of Undamped Electric Oscillations from a Continuous Current.—The discovery that it is possible to produce undamped electric oscillations from a continuous electric current by means of the electric arc opened up a wide field of research. In 1892 Elihu Thomson patented in the United States (U. S. Patent, No. 500,630. Applied for July 18, 1892) the following method intended to effect the above-mentioned transformation. From the terminals of a direct current dynamo or a storage battery B, having an electromotive force of 500 volts, a circuit is taken which passes through a coil of very high inductance K and is interrupted by a spark gap S between two metal balls. These balls are adjustable as to distance, and are also connected by another circuit consisting of a condenser C and an inductance L in series (see Fig. 8).

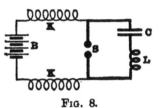

FIG. 8.

The operation was stated to be as follows :—When the spark balls are put in contact, a current is drawn from the supply and passes through the large inductance coil. If the balls are separated, an electric arc is formed and the condenser becomes charged by the difference of potential between the balls. The formation of the arc between the balls involves, however, the passage of a current through the large inductance, which causes a drop in voltage, so that the potential difference of the balls is decreased. The inventor stated that the ball distance, inductance, and capacity can be so adjusted that the condenser is regularly charged and discharged across the spark gap. The electromotive force in the direct current circuit charges the condenser and then forms an arc across the spark gap, but the rush of current which then ensues through the large inductance causes an arc between the balls and a drop in their potential difference, and the condenser then discharges back across the gap. In his specification Elihu Thomson says that he was able easily to obtain *persistent* oscillations in the condenser circuit of 30,000 or 40,000 per second, but no proof was given in this publication that the oscillations were really unintermittent. Nevertheless, it is clear that he realised the utility of undamped oscillations, and was endeavouring to find means for producing them, as shown by remarks made subsequently in 1899 on the matter in an address to the American Association for the Advancement of Science (see the *Electrician*,

September 22, 1899, p. 778). He does not mention the use of any other material than metal for the balls, but it was affirmed that the effect is improved by the use of a strong magnetic field across the arc, or an air blast applied to the space between the balls. These observations of Elihu Thomson did not at the date of first publication attract much attention, probably because no apparent immediate application presented itself, and it was not until a fresh discovery was made by Duddell that interest in the matter revived. It is clear, however, that Elihu Thomson had proved in 1892 that the shunting of a direct current arc by an oscillatory circuit containing capacity and inductance provided a means for converting some of the energy of a direct current into energy of electric oscillations, whether the transformation was into groups of intermittent damped oscillations or into true persistent oscillations.

3. **Duddell's Singing Arc.**—In 1900 W. Duddell described some very interesting observations on the behaviour of an electric arc

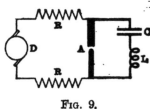

FIG. 9.

between solid carbon terminals when shunted with a condenser and inductance in series, in a paper to the Institution of Electrical Engineers of London, entitled, "On Rapid Variations in the Current Through the Direct Current Arc." In these experiments he formed an electric arc A between rods of solid carbon of the kind generally used as the negative rod in an ordinary plain continuous current arc, but he connected the two arc carbons by an oscillatory circuit consisting of a condenser C and inductance L in series (see Fig. 9). Using carbons 9 mm. in diameter, with an arc current of 3·5 amperes and a potential difference of 42 volts, and a condenser of 1 to 5 microfarads capacity in series, with an inductance of 5 millihenrys, he found that the electric arc gave forth a musical note, the pitch of which depended upon the capacity and inductance in the oscillatory circuit. Also that in the condenser circuit undamped electric oscillations were set up.

He showed that the effect could only be well produced with solid carbons, and this only when the capacity in the shunt circuit was of the order of a microfarad, and the resistance of that circuit rather small. He also noticed that to obtain the effect, the arc must be supplied with continuous current from some steady source, such as a dynamo D or storage battery, and a resistance R of several ohms must be put in this circuit in series with the arc. The resistance of the inductance in series with the condenser must, on the other hand, be low, not more than about 1 ohm.

To explain the manner in which these oscillations are set up in the condenser circuit by the continuous current passing through the arc, we must consider some of the properties of the carbon arc itself, that is, of the electric arc taken between hard carbon rods.

In the case of every conductor of electricity there is some relation between the current flowing through it and the potential difference between the ends of the conductor. Thus if we take a metallic conductor and keep it at a constant temperature, and create various potential differences, reckoned in volts, between its ends, and measure the resulting current flowing through the conductor in amperes, we find that the current is strictly proportional to the terminal difference of potential, provided there is no internal source of electromotive force.

Hence the relation of current to voltage for various current values can be represented by a straight line, as drawn in Fig. 10, in which the abscissæ represent current in amperes and the ordinates the potential difference of the ends in volts, and the tangent of the angle of slope of the line, or the ratio of voltage to current is, by Ohm's law, constant and equal to the resistance of the conductor. Any line

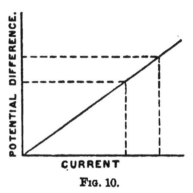

Fig. 10.

representing the relation of the potential difference of the terminals of a conductor or a generator to the current flowing through it or out of it is now called a *characteristic curve*. Hence, for ordinary metallic conductors, the characteristic curve is a straight line rising upwards with increasing current, and is called a *straight rising characteristic curve*. If, however, a series of observations are made by means of a voltmeter and ammeter on an electric arc between carbon rods, we find a totally different form of characteristic curve. If we measure the current through the arc and the potential difference of the carbons for various constant lengths of arc, and plot a curve showing the relation between the two for various currents through the arc, we obtain a curve which is concave upwards and slopes downward as the current increases ; in other words, we have a *falling characteristic* (see Fig. 11). An increase in the arc current is accompanied by a decrease in the potential difference of the carbons. Hence the arc considered as a conductor differs essentially from a metallic conductor, and does not obey Ohm's law.

The characteristic curve of the arc is therefore a curve which slopes in the opposite direction to that of conductors which do obey Ohm's law. Moreover, H. T. Simon has furthermore distinguished between the so-called *static characteristic* and the *dynamic characteristic* of the arc. The former is a curve which delineates the relation between the arc current and the arc electrode potential difference (P.D.) when these quantities are slowly varied and in one direction, and the latter is a curve which delineates the relation between current and P.D. when these quantities vary periodically and in a cyclical manner, as in the alternating current arc.

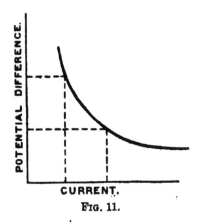

CURRENT.

FIG. 11.

The static characteristics of carbon arcs have been determined and described for many different arc lengths and carbon sizes and qualities, particularly by Mrs. Ayrton. The static characteristics for arcs between carbon and metal electrodes and in various gases have been studied by Upson and others. In all cases they are curves sloping downwards with varying curvature indicating that as the arc current increases the arc P.D. diminishes (see Fig. 12). On the other hand, the dynamic characteristic of the arc is a closed loop like the hysteresis or cyclical magnetisation curve of iron.

When the current and voltage vary periodically, the arc P.D. corresponding to a decreasing current of any value is lower than for the same current when it is increasing. Hence the P.D. is not merely a function of the current, but of the direction in which the current is changing. If we then consider the operations which take place in the carbon arc when so shunted by a condenser and inductance in the light of the above facts, it will be seen that they are as follows :—

Suppose the arc to be burning steadily, and that the condenser shunt circuit is suddenly applied to the carbons. Electricity rushes into the condenser, and the current through the arc is momentarily diminished. The potential difference (P.D.) of the carbons is thereby increased, and thus tends still further to charge the condenser. When the condenser is fully charged, the arc current again slightly increases, and this is accompanied by a small fall in the P.D. of the carbons. The condenser then begins to discharge current through the arc, and still more increase the arc current, and therefore lowers the carbon P.D. Owing to the inductance in series with the condenser, it not only completely discharges itself, but more, it becomes charged in the opposite direction. It is then in a condition to repeat the process with even better conditions, and in this manner persistent electric oscillations are set up in the shunt circuit, the condenser alternately drawing current from the arc and then giving it back again, and owing to the falling characteristic of the arc the accompanying changes of carbon P.D. are such as to sustain the operation. The process may be described as the electrical equivalent of the action of a closed organ pipe. In this last case we have a steady blast of air from the bellows blown in at the foot of the pipe. This corresponds to the direct current through the arc. This air is made to impinge against the lip of the pipe and create a sudden compression in the pipe near the lip. This compressed aerial region is propagated up the pipe, reflected at the top and returns to the mouth. The air pressure then being rather greater inside the pipe than externally, the impinging air jet is forced outside the lip and tends to produce a reduction of pressure or rarefaction inside the pipe near the mouth. This rarefied region is in turn propagated up the pipe and returns again, and on reaching the mouth of the pipe causes the jet of air to play inside the lip, and thus reproduce the wave of compression. In this manner the fluctuating change of air pressure in the pipe controls the behaviour of the issuing air jet, so that it tends to maintain that varying condition and establishes regular periodic changes of air pressure in the pipe and outside it, which are appreciated by the ear as a musical sound. Some part of the energy of the steady air jet is thus converted into aerial oscillations. In the electrical experiment the condenser and inductance constitute an oscillatory circuit which has a natural time period of electric oscillation of its own. It answers to the column of air in the organ pipe, which also has its own natural time of aerial vibration depending on the length of the pipe. The changes of air pressure just outside the mouth of the organ pipe correspond to the changes in potential difference of the carbons between which the arc is formed. The air jet supplied by the organ bellows corresponds to the continuous current supplied to the arc. The organ pipe therefore converts some of the kinetic energy of this issuing jet into energy of aerial oscillations in the pipe. The condenser and inductance shunted across a direct current arc likewise convert some of the energy of the direct current supplied to the arc into energy of electrical oscillations in the condenser circuit. We have furthermore to notice that if the condenser circuit is radiating, that is, giving up energy to the surrounding dielectric, it must be drawing that energy from the arc. Hence, the power taken up by the condenser from the arc when charging must be greater than the power given back by it to the arc when discharging. It follows that the path of the characteristic curve of the arc when its current is increasing must be different from the path when the arc current is decreasing; and, in fact, the complete cycle must form a closed loop. For the characteristic curve is a curve connecting the two variables, current and P.D., which are the factors of power, and the diagram drawn in terms of these is like the indicator diagram of an engine, which is a curve connecting pressure and volume of the working substance. The area of the closed indicator diagram is a measure of the work done by the steam and the area of the closed characteristic curve is a measure of the work done on the condenser circuit.

To have any oscillations produced at all in a condenser shunt circuit we must, therefore, work between two points on the arc characteristic curve, at which it is not flat but sloped, and the steeper the curve in this region the greater will be the power taken up by a condenser of given capacity for a fixed current variation through the arc.

If, therefore, the characteristic curve is not very steep at the point at which we are working, it is necessary to employ a condenser of somewhat large capacity if we are to draw from the arc any appreciable power at each charge, because the voltage variation which creates the charge is small. Under these conditions the time period of oscillation cannot be very small. When using two solid carbons as arc electrodes the characteristic curves for currents up to about 8 or 10 amperes are not very steep, as shown in Fig. 12, for various arc lengths. Duddell found that for such arcs it was necessary to employ a condenser having a capacity of the order of 1 microfarad or more to obtain oscillations of any sensible energy, and that it was not easy to obtain oscillations of a frequency higher than 10,000 ~. Other physicists, who, no doubt, used smaller arc currents and worked, therefore, on the steeper part of the characteristic, were able to use smaller capacities and therefore obtained higher frequencies. Banti, Wertheim-Salomonson, and Maisel found it possible to obtain oscillations having a frequency of up to 400,000 by employing a carbon arc operated with a small current and large voltage. In all probability this was due to the use of smaller arc currents, enabling a smaller capacity to be employed, and yet charged with appreciable energy by working at a steeper part of the characteristic curve. Meanwhile it may be said that the method generally was not regarded, prior to 1903, as affording a very promising means of obtaining powerful undamped oscillations of the high frequency and energy required in radio-telegraphy. Nevertheless, Duddell had foreseen its application for this purpose, and specially mentioned it in a British Patent Specification, No. 21629, of 1900.

This suggestion, however, lay dormant, but was revived again by R. A. Fessenden, who patented in 1902 and 1903 in the United States plans for effecting radiotelegraphy by setting up oscillations in a circuit consisting of a condenser and inductance connected in

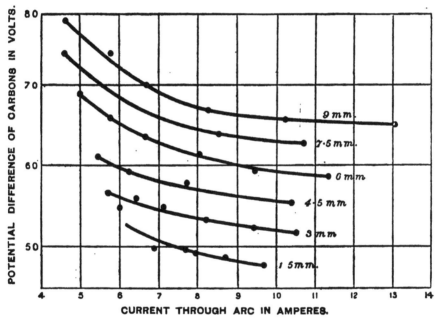

Fig. 12.—The numbers on the curves denote the arc length in millimetres.

series with a spark gap; the said spark gap being also connected through a high resistance with the terminals of a continuous current dynamo. The specification taken alone furnishes no proof that persistent oscillations were produced and not a series of inter-mittent condenser discharges caused by the drop in voltage resulting from the rush of current through a high resistance. Hence, in the absence of definite proof that undamped and un-interrupted oscillations were obtained by this method, we have a right to suspend judgment upon the result.

4. **Poulsen's Method of Producing Undamped High Frequency**

H

Oscillations.—In 1903 Valdemar Poulsen, of Copenhagen, described important improvements in the arc method of creating electric oscillations which give fresh importance to the matter. He produced an electric arc between a carbon rod as the negative and a copper rod as the positive terminal, the latter being kept cool by a water circulation within it. The arc was at the same time surrounded by an atmosphere of hydrogen or a hydrocarbon gas or vapour, and crossed transversely by a strong magnetic field. On shunting this arc with an oscillation circuit consisting of a small capacity and a large inductance, he found that he obtained in this circuit very powerful undamped or persistent oscillations, the

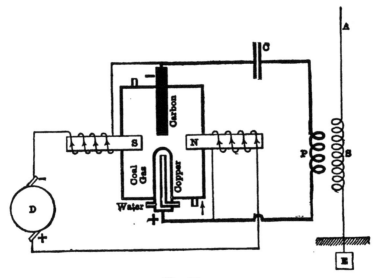

Fig. 13.

frequency of which, by a proper selection of capacity and inductance, could be made to be as high as a million or more, and quite within the range of those required for radiotelegraphic work.

Before proceeding to explain the reasons for the increase in frequency obtainable by the above means, it will be best to describe more in detail the construction of Poulsen's apparatus, and some of the modifications of it which have been suggested.

The electric arc is formed with a direct current voltage of 400 to 500 volts between the end of a solid carbon rod, about 1 inch in diameter, and the end of a water-cooled copper pole (see Fig. 13). The latter consists of a copper tube, which is closed at both ends and has an inlet and exit pipe for circulating cold water through it. The end of this tube terminates in a sharp copper nose piece, which is removable and can be renewed when burnt away. These two electrodes pass through holes in brass sockets let into two marble slabs, which form the ends of a brass cylinder, which passes through a brass box shown in section in Fig. 13. By this means the cylinder can be cooled outside by water to remove the heat created by the electric arc formed between the copper and carbon poles in its interior. Ar-

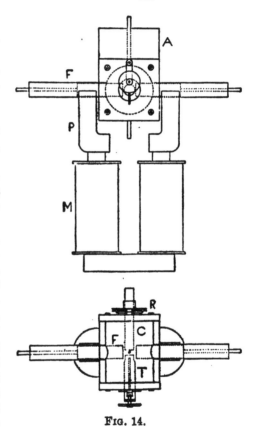

rangements are made whereby the carbon rod can be slowly rotated on its axis by a motor, and the arc is "struck" and length regulated by a screw attached to the copper pole.

In addition to this, the polar ends of a powerful electromagnet M (see Fig. 14) project into the cylinder gas tight so as to form a powerful magnetic field transversely to the arc. By means of a pipe placed under the arc, coal gas can be admitted to the cylinder, and the gas passes out through an exit pipe. When the cylinder is full of gas, the arc is formed by means of the current taken from a 500-volt continuous current dynamo, and the magnetic field excited. The copper pole must be the positive pole and the carbon the negative. If,

FIG. 14.

then, an oscillatory circuit, consisting of a condenser of small capacity in series with an inductance of such magnitude as to give to the oscillatory circuit a natural frequency of 500,000 to 1,000,000, has its ends connected to the copper and carbon rods, powerful electrical oscillations are set up in the condenser circuit. The conditions for obtaining this effect are, however, as follows: Choking coils or inductances must be placed in the wires bringing the continuous current to the arc, so as to prevent the oscillations passing back through the generator. The gas or other hydrocarbon must be supplied freely but not too fast. The magnetic field must be strong, and the arc length must be adjusted to it. There is a particular arc length (called the active arc) which gives the best results. The copper terminal must be kept as cold as possible, and the carbon rod must have its edge square and sharp and be kept in slow but very regular rotation. When these conditions are all fulfilled, the oscillations in the condenser circuit are powerful and practically undamped or persistent.

FIG. 15.

[*Reproduced from the " The Electrician " by permission of the proprietors.*

In place of coal gas, pure hydrogen gas may be used, but it is rather more difficult to maintain a steady electric arc in an atmosphere of pure hydrogen than in coal gas or in air. We may also use the vapour of a volatile hydrocarbon liquid such as pentane, or petrol, or even alcohol may be introduced drop by drop into the arc chamber and allowed to evaporate. This liquid can be supplied from a sight-feed lubricator as shown in Fig. 15, which gives a general view of the Poulsen Arc apparatus. It is found that almost any gas which does not contain oxygen will exalt the frequency of the oscillations obtainable from a carbon-copper arc, even an inert gas like nitrogen, but a hydrocarbon gas or vapour is found to give the best effect. Poulsen proceeded immediately on the discovery of these facts to apply them in the practice of radiotelegraphy, but we shall return to the consideration of this application in a later chapter.

5. **Other Researches on the Transformation of Continuous Currents into Electric Oscillations.**—In addition to the researches already mentioned made on the singing arc, the effect has been the subject of experiments by Simon and Reich, and researches by H. Th. Simon, and the last-named investigator has done much to elucidate the matter as well as to explain the reasons for the discrepancies between the results of other observers.

Simon and Reich in 1903 found that when a high potential arc was formed between metal balls in vacuo, strong oscillations were set up in a circuit including a capacity and inductance in series placed as a shunt across the balls. H. Th. Simon in 1906 subjected the whole phenomena to very careful scrutiny. He pointed out, as already explained, that the production of oscillations by a continuous current arc essentially depends upon the arc having, as already explained, a falling characteristic curve, and showed that the effect was best produced by using high potential arcs and small currents and electrodes which are good conductors of heat like metals.

The reason for this is clear when we examine the form of a characteristic curve of a carbon arc for constant arc length, but varying current as in Fig. 12. We see that the slope of the curve continually increases as the current decreases, so that for small currents the characteristic is much steeper than for large ones. This implies that for small currents a certain absolute decrease in the current is accompanied by a much greater increase in the P.D. of the electrodes than is the case for larger currents. Hence we can communicate the same energy to a condenser of small capacity by an arc using small arc currents as to one of larger capacity by using larger arc currents. In the former case, however, we have a higher frequency of oscillation made possible.

As long, therefore, as we are using a large current to form a carbon arc in air, say, 10 amperes, we are working on the nearly flat part of the characteristic curve, and we can only transform a sensible part of the continuous current energy into oscillations by making use of a condenser of relatively large capacity, and therefore permitting only relatively low frequency oscillations to be created. If, however, we employ a high potential arc and a small arc current, then we are working on the steep part of the characteristic curve, and can transform a sensible portion of the direct current energy into electric oscillations by using a small capacity in the shunt circuit, and hence obtain oscillations of high frequency. This, then, probably explains the discrepancy between the results of various observers. They have used different arc currents and voltages.

The problem of creating undamped oscillations of a frequency high enough for radiotelegraphy resolves itself, therefore, into the discovery of methods for making the characteristic curve of the arc steep for that part corresponding to the current used.

It appears that this can be achieved to some extent by the use of artificially cooled metallic electrodes, or at least a metallic anode or positive pole, and partly by causing the arc to be formed in a strong transverse magnetic field.

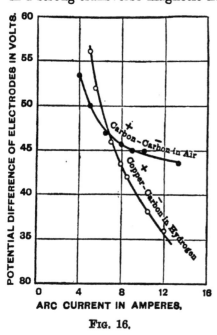

FIG. 16.

Poulsen's important discovery, however, was the effect of a hydrogen or hydrocarbon atmosphere in increasing the steepness of the characteristic curve of the copper-carbon arc. This was well shown by some experiments carried out in the Pender Electrical Laboratory by Mr. W. L. Upson, in 1906 and 1907, at the author's suggestion. In these experiments, amongst other results, the characteristic curves were obtained for electric arcs formed between electrodes of different materials in air and in hydrogen gas. Very considerable differences were found in the form of curves. Thus, for instance, the diagram in Fig. 16 shows the forms of the characteristic curves of a carbon-carbon arc in air, and of a copper-carbon arc in hydrogen, the arc in the last case being formed between a cooled copper rod as the positive and a carbon rod as the negative electrode. It will be seen that for the same arc current the copper-carbon arc in hydrogen has a much steeper characteristic curve than the carbon arc in air. Also the above curves show the importance of using small arc currents and high potentials.

Upson found that a carbon-aluminium arc in air, the carbon being the positive electrode, had almost as steep a characteristic for small currents as a copper-carbon arc in hydrogen, copper being the positive electrode. No explanation has yet been given why the hydrocarbon atmosphere so greatly increases the steepness of the characteristic curve of the arc burning in it. Another matter of considerable interest is whether the oscillations set up by the Poulsen arc are truly persistent or are intermittent. The author examined this question as follows :—

A Poulsen arc apparatus was constructed as described in the previous section, and an oscillatory circuit was connected to the arc electrodes. This consisted of a condenser, C (see Fig. 17), made of metal plates separated by ebonite sheets, the whole being immersed in highly insulating oil, joined in series with a copper wire spiral inductance L. The condenser had a capacity of 0·0029 microfarad, and the inductance was 215 microhenrys. Hence the oscillation period of the circuit was nearly 5 microseconds, or $\frac{5}{10^6}$ of a second. To this oscillation circuit was connected a long helix H of silk-covered wire, wound on an ebonite rod 4·78

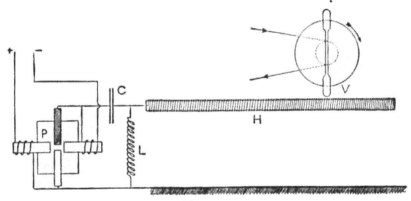

Fig. 17.

cms. in diameter, the helix having 5470 turns of No. 30 S.W.G. copper wire, and a length of 210 cms. Also a strip of zinc as long as the helix was laid on the table, the helix and the strip being connected to the two ends of the inductance of the oscillatory circuit, as shown in Fig. 17.

When the Poulsen arc P was set in operation, the condenser circuit had oscillations set up in it, and these communicated to one end of the helix a series of electromotive impulses. Such a wire helix possesses inductance and capacity, and periodic electromotive forces acting on one end of it set up therefore a condition of electrical vibration in it. The helix has a certain natural period of electrical oscillation of its own, and if the frequency of the electromotive impulses agrees with this period, then the variations of potential at the far or open end of the helix will become very much greater than the variations at the end attached to the condenser circuit. This is a case of electric resonance, and the effect is exactly analogous to that by which the sound of a tuning fork is greatly exalted if it is held over a tall glass jar of suitable depth. If the jar has a depth about equal to one-quarter of the length of the air wave corresponding to the note emitted by the fork, then the conditions are fulfilled for producing this exaltation of the sound. The jar, or column of air in it, is then said to be in resonance with the fork.

In the same manner a helix of wire of suitable length may be brought into electric resonance with an oscillatory circuit. In the experiments described, the far end of the helix of wire was furnished with a number of needle points. The arc was formed by the currents taken from 440 volt supply mains, but some part of this voltage was dropped in the regulating resistance of the arc. The actual potential difference of the arc electrodes (copper and carbon) was from 300 to 350 volts continuous, and the arc current from 5 to 10 amperes. The potential difference of the condenser of capacity 0·0029 mfd. in the shunt circuit was, however, as much as 1200 or 1500 volts R.M.S. value, and the current in this condenser circuit was about 5 amperes, as measured by a hot wire ammeter. This showed that the amplitude of the potential variations of the condenser plates was very much greater than the steady potential difference of the arc electrodes. The helix, however, effected a further rise in the potential, for from the needles at the far end powerful electric brush discharges took place, which showed that the voltage variations at the open end of the helix vastly exceeded even those of the condenser plates. The helix produces a powerful electric field all around it, and in this field vacuum tubes of all kinds, or glass bulbs filled with rarefied gases, glowed brilliantly. A vacuum tube V of the spectrum type (see Fig. 17) filled with rarefied Neon gas glows with an intense orange light when held near the helix when the arc is in operation. If the tube is waved rapidly to and fro, or attached to a turn table and rotated (see Fig. 17), the persistence of vision causes its image to be expanded into a band or disc of light. This image will, however, be found to be crossed by black lines transversely, for the to and fro movement, and radially for the rotation, and this indicates that the light of the tube is intermittently extinguished. The tube is not continuously luminous because the electric field and the electric oscillations are not absolutely uninterrupted. This experiment shows that although undamped oscillations are set up in the condenser circuit, they are not quite without interruption or discontinuity. The cause of these interruptions in the oscillations seems to be the sudden shifting of the point on the carbon from which the arc takes its departure as the carbon rod rotates. On the other hand, if the carbon does not rotate it is rapidly worn away at one point and changes in arc length occur.

Nevertheless, whilst the oscillations are taking place they are undamped in the sense that their amplitude is maintained. In the chapter on radiotelegraphic measurements (Chapter VIII.) we shall describe the methods in use for measuring the damping and logarithmic decrements of electric oscillations. Measurements have been made of the logarithmic decrement of the oscillations set up by the Poulsen arc by Rausch von Traubenberg, and he has found that the log. dec. is practically zero. This implies that the amplitude of each oscillation is the same as that of the preceding or following one, in other words, that the oscillations are persistent. Nevertheless, we must conclude from the author's experiments that under some circumstances short interruptions in the uniform flow of the oscillations may take place. The Poulsen method may, however, be properly described as a method for the production of undamped oscillations. The claims that have been made for other arc or spark methods, such as that of Elihu Thomson, or the modification of it suggested by Fessenden or S. G. Brown, have not yet been justi-

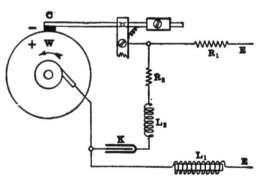

[*Reproduced by permission from " The Electrician."*

Fig. 18.

fied to the same extent, by measurement of the logarithmic decrement, or established themselves by definite evidence as proved methods for the production of persistent oscillations.

S. G. Brown, in 1906, devised a modification of Elihu Thomson's method, as follows : A disc of metal W, preferably of aluminium, is fixed to a shaft and kept in slow rotation (see Fig. 18). Against the edge of this disc a copper block C rests, pressing lightly, and a direct current under a pressure of about 200 volts is passed through a resistance R_1 and large inductance L_1 and across the loose contact between the block and the disc. A condenser K and small inductance L_2 in series are also joined as a shunt between the block and the disc. When the direct current passes, oscillations of high frequency are set up in this condenser circuit, and these can be transformed up or down by an oscillation transformer. We cannot, however, conclude without proof that this method produces persistent oscillations and not a very rapid series of intermittent oscillations. The only convincing evidence that in any particular method provides a means for the production of truly persistent undamped oscillations, is afforded when an actual measurement of the logarithmic decrement shows it to have a zero value. In addition, some evidence should be forthcoming to show that there are no interruptions in the series of oscillations.

Various modifications have been suggested in the actual apparatus for the production of undamped oscillations by the electric arc in hydrocarbon vapour. The author has made use of the following arrangement: To get rid of the necessity for rotating the carbon, the magnetic field itself may be made a radial or conical field, and the arc caused to take place transversely to it. The arc itself will then rotate round the edge of the carbon. To obviate the necessity for a closed box to contain the gas enveloping the arc, the anode or cooled copper terminal may be made cuplike in form at the upper end, and the carbon or cathode may be made hollow, and various liquid hydrocarbons allowed to drop down it so as to generate a hydrocarbon vapour just where it is required. Accordingly the arrangement of the apparatus is as follows :—

On a cast-iron base-plate is placed a cylinder of iron 15 cms. in diameter, and 15 cms. high, and about 2 cms. in thickness. Inside this cylinder is a vertical iron pin, which is enclosed in a brass tube, and cold water can pass up a hole bored in the pin and down again in the space between the pin and the brass cylinder. The top of the brass tube is closed by a recessed copper cap, which is kept cool by the circulating water. This tube is surrounded by a magnetising coil, and when a current passes through this coil it magnetises the pin and creates a flux, which passes up the pin, then spreads out radially and completes its magnetic circuit through an iron disc with a hole in it placed on the top of the cylinder. Through this hole passes a hollow carbon rod, and the current to form the arc passes up the brass tube through the cap, leaping across at some point to the carbon, and thus forms an arc between the carbon and the inner edge of the copper cap, which is constructed with a replaceable ring of copper for renewals. This arc is formed in a strong radial magnetic field, and therefore tends to rotate round the carbon rapidly. Some suitable liquid, such as turpentine, petrol, pentane, or amyl alcohol, is allowed to trickle down the hollow carbon and drop into a copper cap and be volatilised, and thus surrounds the arc with the necessary non-oxygenic atmosphere. To protect the upper iron disc of the cylindrical tubular electromagnet from being overheated, a sleeve of insulating fire-proof material, such as porcelain or silica, is placed round the carbon. The coil exciting the magnetic field may be arranged as a shunt coil on the main arc circuit or be placed in series with the arc. If used in the latter way, it obviates the necessity for any other choking coil in series with the arc, but in any case there must be such a coil, to prevent the oscillations excited in a condenser and inductance circuit joined as a shunt across the arc from passing back into the generator circuit. An arrangement of this kind will produce persistent oscillations when the arc is worked with an electromotive force of 220 volts taken from the electric supply mains of an ordinary house service, provided that it is a direct current supply.

As already mentioned, to secure the best results there is a particular adjustment of magnetic field strength, supply of hydrocarbon or non-oxidising gas, and length of arc is necessary, and these can generally only be found by trial and failure.

The oscillatory circuit which is shunted across the arc must be one in which the capacity is small and the inductance large. If we reckon the capacity C in microfarads and the inductance L in microhenrys, then the ratio $\frac{L}{C}$ must be a large number, something of the order of 10,000. In other words, we must keep the capacity small relatively to the inductance. If made large, that is, anything like a considerable fraction of a microfarad, it will be found impossible to keep the arc alight. The condenser robs the arc of so much current at each oscillation that it is extinguished. There can be no doubt that the arc apparatus for producing undamped oscillations is a somewhat troublesome appliance to manipulate when it is desired to obtain a prolonged production of oscillations, and one difficulty is to get rid of the large amount of energy which is dissipated in the form of heat in the arc itself.

The single arc works well up to about 10 amperes and 440 volts, but when it is desired to obtain more energy in an oscillatory circuit, then it is better to employ a number of arcs joined in series rather than attempt to put more current through a single arc.

An easily managed form of arc generator devised by the Author is as follows:—On a metal pot or vessel is placed a metal lid (see Fig. 19). This lid has in it certain holes bushed with ebonite through which pass steel rods. Each of these carries at its lower end a copper cylinder open at the bottom and closed at the top except for a few small holes. The lid also carries a sliding tube, at the bottom of which is fixed a plate. This plate has fixed to it certain insulated holders, in each of which is a length of arc-lamp carbon. The copper tubes fit over these carbons like extinguishers over a candle. Each steel rod carrying the copper tubes is adjustable independently for height, so as to

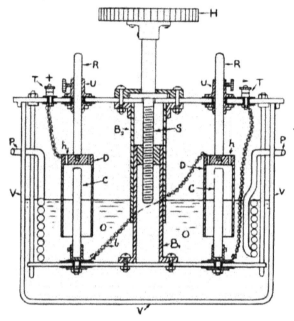

Fig. 19.—Section of Fleming arc generator of undamped oscillations.

make the copper extinguisher just rest on the carbon. Then, by means of a screw, the bottom plate can be lowered and make a space of equal length between the tops of all the carbons and the under side of the tops of all the copper cylinders. By means of insulated wires each copper hat is joined to the carbon of the next pair. If then a voltage is applied we can strike an electric arc between the top of each carbon and the underneath side of its covering copper hat, and all these arcs will be in series. The metal vessel is filled about half full of a heavy mineral oil so that each carbon is immersed in oil up to within an inch of its top and the bottom edge of each copper cylinder is below the level of the oil. The arc soon heats the copper cylinders, and this heat evaporates the oil and fills the closed space in which the arc is burning with oil vapour. This vapour can find an exit through the little holes in the top of the cylinders. The result is to give a series of electric arcs burning in a non-oxygen containing atmosphere. If the whole series of arcs is shunted by a condenser in series with an inductive resistance, and if the arcs are operated with a continuous voltage, we shall have produced in the condenser circuit undamped electric oscillations. We can in this manner work 4 arcs in series off a 220 direct-current voltage or supply, and 8 arcs in series of a 440 volt supply, but it is necessary to add in series with the arcs a little adjustable resistance.

The arrangement requires no magnetic field, but the oil can be kept from rising in temperature beyond a certain point by a coil of lead tube laid round the inner side of the containing vessel through which cold water is circulated.

There are several methods for producing nearly undamped oscillations which may be regarded as intermediate between the alternator and arc methods and the quenched spark methods for the production of intermittent trains of oscillations. These may

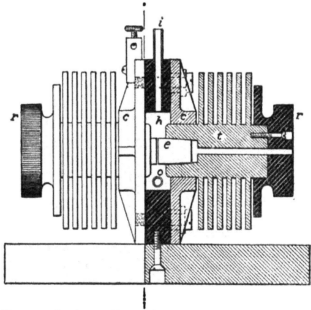

Fig. 20.—Section of Chaffee discharger for producing highly damped discharges.

be described as methods for the production of closely sequent trains of damped waves which run into one another so closely that they are practically equivalent to undamped waves. One of these is due to E. Leon. Chaffee, who has produced such closely sequent trains by means of very rapid condenser discharges taking place between a flat aluminium cathode or negative pole and a copper anode or positive pole.

Two flat surfaces of aluminium and copper are placed a fraction of a millimetre apart in an atmosphere of moist hydrogen (see Fig. 20). If a condenser discharges across this gap, the condenser being maintained constantly charged by a dynamo giving a continuous electromotive force of 500 to 600 volts, there is a very rapid sequence of highly damped discharges.

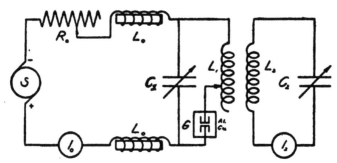

FIG. 21.—Diagram of connections for Chaffee discharger.

If a secondary circuit is connected to the condenser circuit (see Fig. 21), then in this secondary circuit we have a close sequence of feebly damped discharges which are practically equivalent to an undamped oscillation.

Two other methods have been devised which accomplish very much the same thing. Galletti has invented a method, the scheme of connections of which is shown in Fig. 22, in which a number of condensers, C_1, C_2, C_3, etc., each in series, with an inductive resistance, R_1, R_2, R_3, etc., are charged off constant potential mains through a common condenser C_0. Each condenser, when charged, discharges through its own

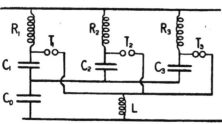

FIG. 22.—Diagram of connections of Galletti multiple discharger.

special gap, T_1, T_2, T_3, etc. The peculiarity of this arrangement is that these discharges do not take place simultaneously but successively, so that in one common inductance, L, there is a closely sequent series of oscillations, and these can be made to create similar secondary oscillations in another syntonic coupled circuit. Galletti states that he can in this manner produce as many as 10,000 sparks per second which follow each other at regular intervals.

Another and perhaps better method of producing a close sequence of spark discharges practically equivalent to an undamped oscillation is due to Marconi. In this case the inventor employs a set of his rotating studded disc dischargers described in Chapter VII. These studded discs are all set on one shaft, but insulated from each other. The discs are each connected with a separate inductance and condenser (see Fig. 23)

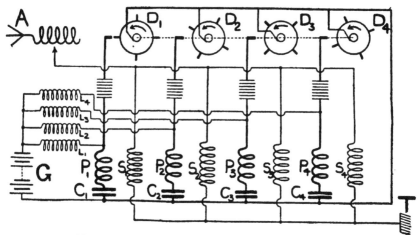

Fig. 23.—Marconi's multiple discharger for the production of undamped oscillations.

so that at intervals a discharge of that condenser is produced. The discs are, however, so set on the shaft that they do not produce these discharges simultaneously but successively. Each separate inductance is inductively coupled to the antenna circuit as shown in the diagram, so that the damped oscillations in each condenser circuit produce inductively secondary oscillations in the antenna circuit which overlap in step with each other as shown in the diagram in Fig. 24.

The result is to produce in the antenna or secondary circuit oscillations which are practically continuous.

Lastly, we may notice a recent extension of the arc method of producing undamped oscillations. There are a certain number of materials, mostly crystalline substances, which have the property that the light contact between them has a unilateral conductivity or conducts electricity better in one direction than the other. The special application of this in the construction of radiotelegraphic receivers is mentioned in Chapter V.

Meanwhile we may mention that it has been found that these materials have not only the property of converting an alternating current into a continuous current when used as detectors or receivers, but have also the property of creating a high frequency current out of a continuous current when used as transmitters.

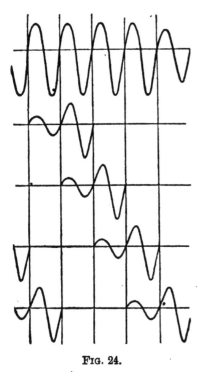

FIG. 24.

The manner in which they are so used is as follows:—If an electric arc is formed with a small current, say, 0·2 ampere and an electromotive force of about 500 to 600 volts, and if a condenser of small capacity in series with an inductance is shunted across this arc, we shall have high frequency oscillations produced in this condenser circuit. It appears that the arc taken between these dissimilar surfaces has a very steep falling characteristic curve, and this, as explained on a previous page, results in the production of oscillations in the condenser circuit.

The substances that can be used as electrodes are as follows:—For the positive electrode aluminium, silicon, ferrosilicon, carborundum, graphite, magnetite, iron or copper pyrites, bornite, zincite, molybdenite. For the negative electrode we can use a metal such as brass, copper, or silver. Thus, for instance, if an arc is formed with a direct current under an electromotive force of 500 volts between small flat surfaces of magnetite (oxide of iron) and brass, the former being the positive electrode, and if this arc is shunted by a condenser circuit, we shall have oscillations set up in the latter. The condenser must have a relatively small capacity, say, 0·02 microforad, and the inductance in series with it must be large, say several thousand or hundred thousand centimetres. This method produces high-frequency undamped oscillations in the condenser circuit, which are useful for such purposes as wireless telephony as explained in the last chapter.

6. Inductive Effects of Undamped Oscillations.—The inductive effects of undamped electric oscillations are very striking. Some of them may be exhibited by the following experiments. A Poulsen arc apparatus is set up as described in the previous sections, and part of the inductance in the oscillatory circuit is made to consist of a coil of insulated wire of 8 to 12 turns, say, of 7/20 S.W.G. indiarubber-covered wire, wound on a square wooden frame, say 60 cms. or 2 feet in the side. If the capacity in the oscillatory circuit is a condenser of 0·003 of a microfarad, and the total inductance is about 300 microhenrys, then the natural time period of oscillation of this circuit would be 5·96 or nearly 6 microseconds, and if the arc were operated with 400 to 500 volts, and taking 7 to 8 amperes, a hot wire ammeter placed in the oscillatory or shunt circuit would indicate a current of 5 or 6 amperes. Hence, although the condenser has a small capacity, it is charged and discharged so many times per second, viz. $\dfrac{10^6}{6}$ or nearly 166,000, that the actual quantity of electricity passing and repassing each section of the conductor is very large.

It must be remembered that an electric current is measured by the quantity of electricity passing any section of its conductor per second, and, as regards heating effects, it does not matter whether this passage is uniformly in one direction or an ebb and flow backwards and forwards. Hence, a small quantity of electricity oscillating very rapidly may produce the same thermal effects as a larger quantity oscillating more slowly. If the quantity of electricity in, or the charge of, a condenser varies from instant to instant, in accordance with a sine function of the time, so that the actual charge q in the condenser is to the maximum charge Q in the relation $q = Q \sin pt$, where p stands for $2\pi n$ and n is frequency of the oscillations, then the current a flowing into or out of the condenser at any instant is the time rate of change of the charge, and is given by the expression $a = pQ \cos pt$.

Hence, in accordance with the principles already explained in Chapter I., the maximum value of the current, which we will denote by A, is connected with the maximum value of the charge Q by the relation $A = pQ$. If the condenser has a capacity C microfarads, and is charged to a potential V volts, then the maximum value of the current into or out of it is A amperes, such that

$$A = \frac{pCV}{10^6} = 2\pi n \frac{CV}{10^6}$$

Accordingly, if the capacity C is, say, 0·003 mfd. and the charging voltage 1500, and the frequency n is $\dfrac{10^6}{6}$, then A should be 4·7 amperes nearly. This calculation proceeds on the assumption that the charge of the condenser varies in accordance with a sine law. This, however, is not strictly true in the case of the oscillatory circuit of a Poulsen arc. Therefore the relation between the current A in the oscillatory circuit and the oscillatory potential difference of the condenser terminals is not strictly, but only approximately, in accordance with the equation

$$A = \left(\frac{2\pi n C}{10^6}\right)V$$

We must note in passing how the true potential difference of the condenser terminals denoted by V in the above equation is to be measured. If we apply a direct current voltmeter to the electrodes of the Poulsen arc, we should obtain a reading V_0, which might be anything between 200 to 500 volts, according to the electromotive force of the direct current dynamo supplying the current and the controlling resistance in series with the arc. If we apply an electrostatic voltmeter to the terminals of the condenser in the shunt circuit, we should find a much larger potential difference, say V_1, which is alternating. But this observed potential difference V_1 is due partly to the direct P.D. between the arc terminals and partly to the oscillations in the shunt circuit, and the true oscillatory potential difference of the condenser terminals, denoted by V in the equation above, is connected with V_0 and V_1 by the relation

$$V_1^2 = V_0^2 + V^2$$

Hence
$$V = \sqrt{V_1^2 - V_0^2}$$

For the actual variation of the P.D. of the condenser terminals may be represented by the ordinates of a line which is made up of a sine curve, the ordinates of which are all increased by a constant amount V_0, and it is not difficult to show that the mean of the squares of the ordinates of such a curve is equal to the sum of the square of the constant ordinate and the mean square value of the ordinate of the sine curve taken alone. To obtain therefore the value of V suitable for insertion in the equation $A = \dfrac{2\pi n C}{10^6}V$ for giving the current, we have to take the square root of the difference of the squares of the readings of an electrostatic voltmeter applied to the condenser terminals and a direct current voltmeter applied to the arc electrodes.

The power taken up in the arc in watts cannot be precisely measured by the product of the direct current through the arc measured in amperes, and the reading V_0 of the direct current voltmeter. It can only be properly measured by a wattmeter. Nevertheless, experiment shows that the *power factor* of the Poulsen arc, viz. the ratio of the true power taken up by it as measured by a wattmeter to the ampere-volts or product of the current and arc P.D., is not far from unity, generally about 0·97.

Returning then to our experiments; if we hold near to the above-mentioned square circuit another circuit consisting of a few dozen turns of highly insulated wire, say, 50 turns of No. 16 S.W.G. india-rubber covered wire wound into a circular or square coil 30 cms. in diameter, and attach to the ends of this circuit a 50-volt carbon filament glow lamp, we shall find that when this secondary circuit is held near to the square primary circuit the incandescent lamp glows up brilliantly. This shows that in the secondary circuit a very considerable voltage is induced. This electromotive force is created in the secondary or lamp circuit by the rapid change in direction in the lines of magnetic force due to the primary oscillation circuit which are linked with the secondary circuit. The oscillatory current in the primary circuit creates around it a rapidly alternating magnetic field. Some of the lines of this field are thrust through or linked with the secondary circuit, and the insertion or withdrawal of these gives rise to the secondary electromotive force. This last is proportional in magnitude to the rate at which the flux linked with the secondary is changing. If I_1 is the maximum value of the high frequency current in the primary circuit, then we may denote the maximum value of the total number of lines of force due to it which are linked with the secondary by MI_1, where M is a quantity called the mutual inductance of the two circuits. The secondary E.M.F. is then measured by the rate at which the flux linkage with the secondary circuit varies. Hence, the maximum value of the induced secondary electromotive force E_2 is equal to pMI_1 or to $2\pi nMI_1$, where n is the frequency. Accordingly, although M and I may be small, yet if n is very large the secondary E.M.F. may become very great. If this secondary circuit has an effective resistance R_2 and an inductance L_2, then the secondary current created in it has a value I_2 such that

$$I_2 = \frac{MpI_1}{\sqrt{R_2{}^2 + p^2L_2{}^2}}$$

as shown in treatises on alternating currents.[1]

In the case of high frequency circuits the resistance R is nearly always negligible in magnitude compared with the reactance pL, and hence we may say that for such a simple inductive circuit the secondary current I_2 is given by the equation

$$I_2 = \frac{M}{L_2}I_1$$

It is therefore increased by reducing the inductance of the secondary circuit as much as possible, whilst increasing the mutual inductance.

This may be illustrated by the following experiment:—

Construct a circuit consisting of, say, 50 or more turns of highly insulated wire No. 16 S.W.G., the wire being wound in a flat, circular coil about 45 cms. or 18 inches in diameter. Insert this coil in series with a condenser of small capacity, say, 0·003 mfd., as a shunt circuit on a Poulsen arc. Then bend a very thick copper wire, say, of 5 mms. diameter or 0·2 inch in thickness, into a nearly complete circle of 45 cms. or 18 inches diameter and connect the ends by a few inches of thinner copper wire, about No. 22 or 26 S.W.G. Fix the copper ring to a wooden handle. Then excite undamped oscillations in the first-mentioned coil and bring down over it slowly the thick copper secondary circuit. The fine piece of copper wire will be rendered red hot and melted. We have here constructed a pair of circuits, the primary of many turns and the secondary of only one turn, so that the mutual inductance M is large, but the secondary inductance L_2 is small. Hence the secondary current induced is many times greater than the primary current and easily melts part of the secondary circuit.

In this case we have constructed a "step-down" transformer for undamped oscillations, decreasing the voltage but increasing the current. On the other hand, we may reverse the process and construct a step-up transformer. If we insert the primary coil of an ordinary induction coil as part of the inductance in the shunt circuit of a Poulsen arc, we can obtain from the ends of the secondary circuit a flaming discharge which resembles an alternating current arc in being a lambent mobile flame rather than a spark, not unlike the discharge produced by a Wehnelt break.

[1] See "The Alternate Current Transformer," by J. A. Fleming. Vol. I., p. 178.

7. Resonance Effects in connection with Undamped Oscillations.—
All the inductive effects of persistent oscillations are vastly
increased if we avail ourselves of the exalting influence of
resonance.

The oscillatory circuit shunted across the Poulsen arc has a
certain time period of oscillation of its own, determined by its
capacity and inductance. If we cause this circuit to act inductively
upon a secondary circuit which contains no condenser, the effect
on this last circuit is to produce a forced oscillation. If, however,
the secondary circuit has a condenser inserted in it and possesses
inductance, then it also has a natural time period of its own, and
by adjustment of the capacity and inductance can be tuned to the
period of the primary. When this is the case, the electromotive
force produced by the reversal of the direction of the magnetic
flux due to the primary circuit which passes through the secondary
has a cumulative action, each impulse adding its effect to the
previous ones, just as when we apply small blows or puffs of air
to a pendulum, striking or blowing exactly in time with the
natural time period of the pendulum. In this last case small
repeated blows soon create a very large vibration in the pendulum,
and in the electrical case the feeble but repeated inductive
impulses finally create a very much larger current than would be
the case if the circuits were not in tune. These facts may be
illustrated in the following manner. Let the shunt circuit of the
arc consist of a small condenser, having a capacity say of
0·003 mfd. and an inductance of about 300 microhenrys, part of
this inductance consisting of a circuit of 12 turns of insulated wire
wound on a square wooden frame, about 60 cms. or 2 feet in the
side. The inductance in series with this square circuit should
consist of two parallel spirals of copper wire, about 10 turns to the
inch, the spiral being 1 inch in diameter and say 36 inches long.
These spirals are provided with a short-circuiting bar, as already
described in § 7, Chap. II., by means of which the total inductance
can be gradually varied. The secondary circuit should consist of a
similar square coil, sliding spiral inductance and condenser, which
may conveniently be a large Leyden jar. In some part of the
secondary circuit is inserted a small 4-volt carbon filament glow
lamp, taking a current of about 0·5 ampere to render it incandescent (see Fig. 25). The secondary circuit is placed about
6 feet, or say 2 metres, away from the primary circuit, the square
coils having their planes parallel to each other. If, then, oscillations
are set up in the primary circuit by the arc, and the distance
between the coils is sufficient, we shall find that little or no effect
is produced in the secondary circuit, as judged by the emission of

light by the glow lamp, whilst the two circuits are out of tune. It is, however, possible to so adjust the inductance and tune the circuit that the little lamp glows brilliantly, thus indicating that the secondary current is immensely increased by tuning, but is extinguished by a very small alteration in the inductance either of the primary or the secondary circuit.   This experiment illustrates very well the effect of syntony in exalting the secondary current. It can be conducted also with undamped oscillations set up in the primary circuit by the use of a spark gap, but it is found that when using undamped oscillations the tuning, as it is called, is much sharper than with damped oscillations.   The reason for this is generally as follows :—

The current set up in the secondary circuit is determined as to strength by several factors.

Let J stand for the root-mean-square value of the secondary

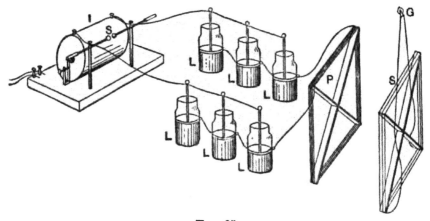

Fig. 25.

current, that is the value which would be indicated by a correct hot-wire ammeter inserted in the secondary circuit.   Then J^2 is called the mean square value of the oscillations, or, by some German writers, the *integral effect* of the oscillations.   Let C_1 be the capacity in the primary circuit and C_2 that in the secondary circuit, and V_1 the potential to which the primary condenser is charged, and let k be the *coefficient of coupling* of the circuits, and furthermore let δ_1 and δ_2 be the logarithmic decrements of the two circuits.   Then when the circuits are adjusted to resonance so that they both have the same time period of oscillation n the secondary current comes to a maximum value.   We shall denote this maximum value by J_{max}, so that J^2_{max} is the mean-square or effective value of the current in the secondary circuit when it is tuned to the primary.

It was shown by P. Drude by an elaborate course of reasoning that the value of this maximum current is given by the formula,

$$J^2{}_{max} = \frac{1}{4}\left(\frac{C_1V_1{}^2}{2}\right)C_2 \cdot \frac{n\pi^4k^2}{\delta_1\delta_2(\delta_1+\delta_2)}$$

The quantity $\frac{C_1V_1{}^2}{2}$ is the energy put into the primary condenser initially or at each charge. Hence it is seen that the secondary current (mean-square value) is proportional to the initial energy given to the primary condenser and to the capacity of the secondary condenser, and to a function which depends upon the coupling and the damping of both circuits. For the proof of this formula, the original paper or advanced treatises on the subject must be consulted. The important point to notice is that the value of $J^2{}_{max}$ is increased by decreasing the log. dec. of either the primary or secondary circuit. This points to the importance of securing small resistance in the circuits. It will be shown in the next chapter that part of this damping depends upon the power of the circuits to radiate their energy, and hence in practical work it is never possible to make the decrements absolutely zero, or else the secondary current would become infinite in value.

The above formula is of use in calculating the current in the antenna of a radiotelegraphic transmitter apparatus when the part which radiates is inductively coupled to the part which stores the energy used.

CHAPTER IV

ELECTROMAGNETIC WAVES

1. **The Electromagnetic Medium.**—Although the study of distance actions in electricity and magnetism, such as the attraction and repulsion of magnetic poles, and of conductors conveying currents, as well as the effects of magnetic and electrical induction, long ago suggested the idea of an electromagnetic medium to the minds of Ampère, Faraday, and Henry, as the means by which these effects at a distance are produced, the notion was hardly more than a surmise until James Clerk Maxwell, in 1864, communicated to the Royal Society a classical memoir on " A Dynamical Theory of the Electromagnetic Field." In this paper he presented the chief facts of electromagnetism in such a manner as to show that electric and magnetic effects cannot be produced instantaneously at a distance, but must be propagated through space with a finite velocity.

The phenomena of optics had already been found to be best explicable on the hypothesis that all space is occupied and all matter interpenetrated by an imponderable medium· capable of undulation, and that waves in it of a certain kind and range of wave length constitute light considered as a physical agent.

Maxwell showed that the velocity with which an electromagnetic effect is propagated through a dielectric is dependent upon the known electric and magnetic qualities of it, and from such measurements as were available at that date, he proved that it must be identical with that of light. Hence, the conception that the assumed luminiferous æther must be identical with the hypothecated electromagnetic medium was advanced to the position of a scientific theory supported by some important evidence.

2. **Electric and Magnetic Quantities.**—To explain the steps by which this conclusion was reached, we must prepare the way by some definitions of terms and special conceptions. In addition to its ordinary state, every material substance can be put into a condition in which it is said to be electrified or charged with electricity.

We are able, for instance, to bring about this state by the friction of two bodies one against the other, but the simplest investigation shows that the resulting electrical condition of the two bodies is not the same, and that whilst both are electrified, the electrifications on them are different in kind. Thus the friction of glass against silk produces so-called positive or vitreous electricity on the glass and negative or resinous electricity on the silk, whilst the friction of ebonite or shellac against flannel produces negative electrification on the ebonite or shellac and positive on the flannel. If the two rubbed bodies are held at a little distance, an attractive force is found to exist between them, and if another small electrified body, say, a pith ball, is placed in the space between the silk and the glass which have been rubbed together, it will tend to move one way or the other, according to the nature of the electrification on the pith ball. The space between two such oppositely electrified bodies is called an *electric field*, and is said to be the seat of *electric strain*, or *displacement*. The force between the electrified bodies is not merely an action at a distance, but is determined by the nature of the material substance, whether solid, liquid, or gas, which occupies the interspace. If a couple of bodies are electrified, one positively and the other negatively, and are found to attract each other with a certain force in air when placed at a certain distance, then if immersed under turpentine or paraffin oil, they would at the same distance only attract each other with about half that force. If, on the other hand, they were placed in a highly perfect vacuum, they would attract each other with a slightly greater force. Hence, we see that the attraction essentially depends upon some quality of the surrounding material medium, but is not entirely dependent on it, for it exists even when all surrounding matter is removed. This quality is capable of being numerically defined, and is called the *dielectric constant* of the medium. We shall denote its magnitude by the letter K. The dielectric constant of an absolute vacuum is taken as unity. We may, therefore, say that the electrified bodies produce a state called an electric strain in the material, then called the *dielectric*, around them, and the degree or intensity of this state is determined by the dielectric constant of that material. It is convenient to regard the state of electric strain in a dielectric as produced by an agency called *electric force*, but determined as to degree or intensity by its dielectric constant. We can then say that electric force produces electric strain in a dielectric to an extent determined by its dielectric constant, just as we say that the bending, flexure, mechanical strain of a beam is produced by mechanical force to an extent determined by the *elasticity* of the beam.

The dielectric constants of some well-known materials have been already given in Chapter I., but owing to the variation in the constant of different specimens of the same substance the numbers given can only be taken as approximations for that of any particular sample.

We can then define a unit of electric quantity or electrification and a unit of electric force as follows: Let two very small spheres be supposed to be equally charged with opposite electricities and placed with their centres 1 cm. apart in a vacuum. If the electric quantity is such that the attractive force is $\frac{1}{981}$ of the weight of a gramme in London, viz. 1 dyne, then each body is said to be charged with 1 *electrostatic unit* of electricity. At a distance of 1 cm. from its centre each such small electrified sphere would exert a unit electric force when placed in vacuo. We can then, by appropriate methods, measure the charge or quantity of electricity on a body in electrostatic units, and also the electric force at any point in an electric field in the above-mentioned units of electric force.

Turning next to the elementary facts of magnetism, we find we can make similar statements. In addition to their ordinary condition, certain bodies, such as steel, can be put into a state in which they are said to be magnetised. If a pair of steel wires are magnetised in the direction of their length, we find that between the ends of these wires, when near together, there are attractive or repulsive forces. The space between these ends and all around the magnetised steel is called a *magnetic field*, and is said to be traversed by *magnetic flux*. A simple experiment with a pair of such magnetised wires suffices to show that the ends are not identical in properties, but differ like positive and negative electricity. Moreover, a very careful experiment would show that as in the case of the electrified bodies so for the magnetised ones, the force between them depends to some extent on the interposed medium. Two long uniformly magnetised steel wires, whose polar ends attract or repel each other at a certain distance with a certain force in air would attract or repel each other with a slightly less force if placed in liquid oxygen or in a solution of ferric chloride, but the force would not vanish even if the magnets were in a perfect vacuum. Hence we are led to conclude that the force depends to some extent upon a quality of the surrounding medium which is called its *magnetic permeability*.

The permeability of empty space is taken as unity, and that of any other substance is denoted by the symbol μ, and is measured by a number greater or less than unity. It is then convenient to consider that this state, called magnetic flux, produced in the space around a magnetised body is due to an agency called *magnetic force*, the degree or intensity of the flux depending upon the permeability of the medium.

Hence, we can say that electrified bodies exert *electric force* (E) in their neighbourhood, and produce *electric strain* (D) in the material or medium by which they are surrounded to a degree depending upon its dielectric constant (K). Similarly, magnetised bodies and also electric currents exert magnetic force (H) in their neighbourhood, and produce magnetic flux (B) in the material or medium by which they are surrounded to a degree depending upon its magnetic permeability μ.

The fact that electrified bodies or magnets attract or repel each other at a distance, and that electric currents can create other currents in wires at a distance, and that these actions are not entirely dependent upon the presence of any material substance in the interspace, but can take place also through a perfect vacuum, has always impressed competent thinkers with the idea that there must be an electromagnetic medium by means of which these actions are transmitted across intervening space.

The observation that light takes time to pass from one place to another, and that it comes to us from far distant stars across interstellar space, which, as far as we know, is not full of ponderable matter, is a proof that it must either be a substance bodily transmitted like a letter sent by post or a physical state or change of state which is propagated through a stationary medium. Innumerable facts of optics prove that it is, in fact, an undulation, and that, therefore, there must be something which undulates. The velocity with which this undulation travels has been measured with considerable exactness, and has been found to be very close to 300,000 kilometres per second, or, roughly, about 1000 million feet per second. The speed with which any disturbance travels through an elastic medium is, as shown further on, determined by the square root of the ratio of its elasticity to its density. No known form of tangible or gravitative material has such a large ratio of elasticity to density as to permit an undulation or impulse of any kind to travel through it at a speed of 300,000 kilometres per second. The atmosphere, for instance, is a material possessing elasticity or resistance to compression, and likewise density. If, however, a sudden compression is created in it at any place, this state of compression is propagated through it at the rate of only about 330 metres per second at 0° Cent., viz. with the velocity of sound. Even in hard steel the velocity of propagation of a compressional or extensional strain travels only at the rate of 5612 metres, or about 18,600 feet per second. Accordingly, we are compelled to admit that if light is due to vibrations propagated through a medium at the rate of. 300,000 kilometres or 186,000 miles per second, the medium capable of this must possess qualities very different from those of any form of tangible or ponderable matter with which we are acquainted. The medium called the æther must necessarily be universally diffused, and must interpenetrate all ordinary matter. It cannot be exhausted or removed from any space, because no material is impervious to it. As far as we know, it is non-gravitative, but ordinary matter stands in some very close relation to it. It must also possess some form of elasticity, that is resistance to a change of state of some kind produced in it, and it must also possess inertia or a quality in virtue of which a change so made in it tends to persist. We are not justified in making the assumption that its elasticity like that of ordinary matter is a resistance to change of bulk or form, or its inertia necessarily an inertia with regard to motion. It is, however, clear that the medium has the power of storing up energy in large quantities and transmitting it from one place to another, as shown by the fact that enormous amounts of energy are transmitted from the sun to the earth. It is only in virtue of some form of elasticity and some type of inertia that a medium can thus transmit energy in the form of undulations through it. When we consider the relations of electric strain, electric force, and dielectric constant, we see that they are quite analogous to the relations which ordinary mechanical stress or force and material strain or displacement and electric resilience bear to each other. The term *strain* in physics means any deformation of a body or change in relative position of its parts, and the word *stress* is applied to denote that which produces strain. If the strain increases with the stress and disappears spontaneously when the stress is removed, the body is called elastic. The ratio of stress to strain at any stage expressed in appropriate units is then called the *elasticity* corresponding to that strain. Thus, gases do not resist change of form but resist change of bulk, and their elasticity is thus a resistance to change in volume, the elasticity being defined as the ratio of the increment in pressure to the decrement in volume produced by it. Solid bodies resist change of form, say extension. Thus a longitudinal stress produces an extension of a bar of metal. The ratio of the pull or tension to the extension, expressed as a fraction of the original length is called the longitudinal elasticity, or Young's modulus of elasticity. The characteristic of an elastic substance is therefore that it experiences some kind of strain under the action of a corresponding stress, and that within certain limits the strain is directly proportional to the stress and inversely as the elasticity.

Moreover, the strain disappears more or less completely as soon as the stress is removed. If, then, we consider a dielectric of any kind we find that a certain state can be produced in it, called electric strain, by an agency called electric force, and that the strain is within limits proportional to the force and to a constant called the dielectric constant. Also, just as the removal of the mechanical stress causes the resulting strain to disappear, so the removal of the electric force causes the dielectric strain to disappear. Accordingly, we notice that the reciprocal of the dielectric constant is analogous to the elasticity of a material substance.

There is more than a mere analogy between mechanical strain of a solid body and dielectric strain. It is well known that the optical properties of transparent bodies are affected by mechanical strain. Thus the mechanical strain produced in the particles of a bar of glass when bent can be rendered evident by examining the bar by polarised light when bands of colour indicate the lines of strain. In the same manner the electric strain set up by electric force in dielectrics affects their optical qualities, and can be rendered evident by the employment of polarised light. Furthermore, in the case of mechanical strain, the production of a strain involves the expenditure of energy, and the strain when produced is a store of potential energy. The energy stored up per unit of volume is measured by the product of the strain and the average stress. Hence it is equal to half the product of the strain and the maximum stress, or to half the quotient of the square of the stress by the elasticity, or to half the product of the elasticity and the square of the strain. In precisely the same manner it can be shown that when a dielectric is in a state of electric strain under electric force, the energy stored up in it per unit of volume is numerically equal to half the product of the square of the electric force and the dielectric constant, or to half the quotient of the square of the electric strain by the dielectric constant.

3. **The Nature of a Wave.**—We have then to consider the production of a wave in an elastic and dense medium. Physically speaking, a wave is defined as a cyclical change taking place in a medium which is periodic in space as well as in time. Put into less formal language, it means that each particle of the medium executes some movement or experiences some change which is repeated over and over again, all particles performing the same motion or experiencing the same change in succession but not all simultaneously. Thus, for instance, a surface wave on water is caused by the particles of water rising and falling periodically, so that along a certain line the particles execute this motion successively. At regular intervals, however, along the line will be found

particles which are in the same phase of their motion at the same instant. These are said to be separated by one *wave length*. The wave length, therefore, is a distance which comprises one set of particles in all possible stages of their periodic motion.

In regarding a wave we may suppose ourselves to remain fixed at a certain point, and to watch the cyclical changes which take place at that point in a time called the *periodic time* (T). Otherwise we may imagine ourselves to travel with a uniform velocity along the line of propagation so as to remain always in contact with the same phase of the motion. This velocity is called the *wave velocity*. Thus a bather standing in the sea fixed at one point finds the water rise and fall over him as the sea waves travel past him, and he can note the interval of time between two successive greatest elevations of the water. But a seagull, flying along over the surface of the sea, by adjusting his speed, can keep himself constantly poised over the summit of a hump of water or place of greatest elevation. His velocity is then that of the wave motion in the direction of his flight.

If we call the wave length λ, and the wave velocity V, and the wave period T, then the relation

$$V = \frac{\lambda}{T} = n\lambda$$

holds good in all cases of wave motion. The wave velocity is equal to the quotient of wave length by periodic time or to the product of wave length and frequency.

On certain assumptions we can obtain an expression for the wave velocity in terms of the elasticity and density of the medium as follows :—

Let us suppose that there are a row of particles each of unit volume and of mass m which lie in a straight line, and let each

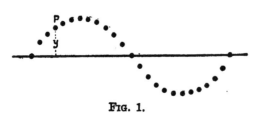

FIG. 1.

particle in succession execute a simple harmonic motion up and down along a line at right angles to the first. A simple harmonic motion is the motion of the projection on any diameter of a point which moves with uniform velocity round a circle. This motion is equivalent to assuming that the small masses lying in one line are attracted back to it with a force varying as the distance when displaced perpendicularly from it. Thus let P (Fig. 1) be one of the particles, and at any moment let its displacement from the zero line be y, and let ey be force drawing it back, then ey is the stress, and the displacement y is the strain, and the elasticity or ratio of stress to strain is e. Then the force acting on the particle of mass m is $m\dfrac{d^2y}{dt^2}$ and this must be equal to $-ey$, because the force tends to reduce the displacement. Hence, the equation of motion is

$$m\frac{d^2y}{dt^2} + ey = 0$$

The above is called a differential equation, and the reader who has some slight knowledge of the differential calculus will be able to see that a particular solution of this equation is

$$y = \mathrm{Y}\sin\sqrt{\frac{e}{m}}.t$$

where Y is the maximum displacement during the phase. This can be at once proved by differentiating the last equation twice with respect to t, and substituting the differential coefficient so obtained in the equation of motion, when it will be found to satisfy it.

Suppose, then, that the particles all execute this motion successively, so that at any one instant their positions delineate a sine curve. If we reckon the abscissæ x from the point on the line at which one particle is on the axis at the zero of time, then the equation to the space distribution of all the particles at any fixed time is $y = \mathrm{Y}\sin x$. Again, suppose that whilst the particles are in oscillation we move uniformly forward so as to keep in contact always with a displacement of constant value y whilst varying our abscissa x. Our velocity will be $\dfrac{x}{t}$, and will be that of the wave motion V. For that ordinate y of constant value we have the relations

$$y = \mathrm{Y}\sin\sqrt{\frac{e}{m}}.t = \mathrm{Y}\sin x$$

Hence
$$\frac{x}{t} = \mathrm{V} = \sqrt{\frac{e}{m}}$$

In other words, the wave velocity is the square root of the quotient of the elasticity by the mass of the particle. If, then, we are considering the oscillations, not of a single row of particles but of elements of volume of a continuous medium, we can write the symbol ρ instead of m, where ρ is the mass of the unit of volume or density of the medium, and we arrive at the formula given above for the wave velocity, viz.

$$V = \sqrt{\frac{e}{\rho}}$$

This expression is the well-known formula for the velocity of a sound wave or wave of compression and rarefaction in any medium, as, for instance, through air, and from it we can determine the wave velocity if we know the elasticity and the density.

Thus, the density of tempered steel is 0·285 pound per cubic inch. The elasticity (Young's modulus) for the same steel is approximately 16,100 tons per square inch. Accordingly, if we reduce this last figure to absolute units in terms of the inch, pound, and second as units, we have

$$e = 16,100 \times 2240 \times 32\cdot2 \times 12$$
and $\qquad \rho = 0.285$

Hence the velocity of propagation of a longitudinal wave of compression and extension in tempered steel is

$$V = \sqrt{\frac{e}{\rho}} = \sqrt{\frac{16100 \times 2240 \times 32\cdot2 \times 12}{0\cdot285}} = 221,000$$

This velocity, however, is in inches per second, but reduced to feet per second it is 18,400, which agrees with the observed velocity of sound through steel. The same expression would enable us to predict the velocity of a wave of any kind, say, a wave of transverse displacement, or shear wave, provided we can obtain the numerical value in absolute measure of the elasticity of the material for that particular kind of strain.

Let us then consider what are the qualities of an electromagnetic medium which would correspond to the elasticity and density of a material substance. We know nothing about the mechanical structure of the æther, nor what forms of strain can be imposed upon it, but we do know that electric force produces in a dielectric and in the æther a state called an electric strain, and that the reciprocal of the dielectric constant is a measure of the ratio of the electric stress to the electric strain, or, in other words, of the electric elasticity. Elasticity is that quality of matter in virtue of which energy can be stored up by strain in it in a potential form. A bent rod, a stretched spring and compressed air, are all cases of strained materials which possess potential energy in virtue of the elastic resilience concerned. If we electrically strain a dielectric we store up in it per unit volume energy equal to $\frac{1}{2}\frac{D^2}{K}$, where D is the electric strain and K the dielectric constant, and this energy is potential and it corresponds to the potential energy stored up in a mechanical form when a material substance of elasticity e experiences a configurational strain S, for the energy so stored up per unit of volume is then $\frac{1}{2}eS^2$.

Again, when an electric current A flows through a circuit embedded in a dielectric, the energy associated with that circuit is measured by $\frac{1}{2}\mu LA^2$, where A is the value of the current in amperes, and L is a constant depending on the form of the circuit and μ is the magnetic permeability of the dielectric. When a material body is in motion its kinetic energy is measured by $\frac{1}{2}\rho BV^2$, where ρ is the mean density of the body, B is its bulk or volume, and V its velocity. We know that an electric current is a form of energy, and there are good reasons for considering that the energy involved is kinetic. Hence, we see that in the expression for the electrokinetic energy the symbol μ, or the magnetic permeability of the medium, takes the place of the symbol ρ or the density of the body in the expression for the motional or kinetic energy of ordinary matter. Accordingly, we have reasons for comparing the quality we call the magnetic permeability of a dielectric with the density of material substances and the reciprocal of the dielectric constant with the elasticity of matter, when we are comparing the electrical with the mechanical qualities. If this is the case, then the expression $\frac{1}{\sqrt{K\mu}}$, which is analogous to the expression $\sqrt{\frac{e}{\rho}}$, should be the expression for the velocity of an electromagnetic wave; but before we can decide this matter we must consider more carefully what is meant by an electromagnetic wave.

4. **An Electromagnetic Wave.**—In order that a wave may be produced in a medium, this last must possess two properties, it must *resist* and *persist*.

Thus, in the case of a surface wave on water, the water surface resists being made unlevel. If at any place it is suddenly depressed, as by dropping a stone into it, a force is brought into play to restore the displaced water to its original level. In so doing the water is set in motion, but in virtue of its inertia when

K

so set in motion it persists in motion, and not only moves until it is back in the original position, but moves beyond it and creates an elevation in place of a depression. Then, again, the force of restitution comes into action to depress the surface again, and so an undulatory motion is set up at that point. Moreover, the water at one point is in close connection with the water around it by cohesion, so that elevation and depression at one place cause the water in the immediate proximity of the initial disturbance to share in the same motion, but to lag behind a little in imitating the motion of the first displaced portion. Hence, to create a wave in a medium we must produce some form of strain which in disappearing as potential energy transforms itself into an equivalent in motional or kinetic energy.

One of the important contributions Clerk Maxwell made to this subject was to explain clearly the manner in which a connection between the potential or electrostatic and kinetic or magnetic forms of energy is established in the case of dielectric media. He pointed out that an electric strain, or displacement, *whilst it is changing*, that is, whilst it is increasing or diminishing, is equivalent to an electric current, and must therefore create magnetic flux along a closed line embracing the varying electric strain. A varying electric strain is therefore called a *displacement current*. Again, Faraday had shown that the variation of magnetic flux through a closed metallic circuit creates electromotive force in that circuit. Maxwell extended this idea also to a circuit in any material, not only conductive but dielectric, and stated that the variation of magnetic flux through any area enclosed by a line drawn in a dielectric must create electric strain along that line. Suppose we have a currrent flowing in a straight infinitely long conductor, magnetic flux is distributed in circular lines round that conductor in the space outside and inside the wire.

If a magnetic pole, that is the end of a very long thin magnet, is held in the field of the current it will tend to rotate round the wire, being urged by a force proportional to the strength of the pole and to the magnetic force at the point where it is held. Experiments proving this are described in every book on physics.

If, however, a short magnet is laid on a disc of card, the card being suspended so as to be free to rotate in a plane perpendicular to the current, but the magnet not free to move on the card, there will be no rotation of the latter (see Fig. 2).

When this experiment is carefully considered, it will be found to prove that the magnetic force (H) at any point near the long straight current must be inversely as the perpendicular distance of the point from the current, and further an analysis shows that it is proportional to twice the current and inversely as the distance.

If, then, we consider a single circular line of magnetic flux of radius r, its length is $2\pi r$, and the magnetic force all along it has a value $\dfrac{2C}{r}$, where C is the current in the wire. The product $2\pi r \times \dfrac{2C}{r} = 4\pi C$ is called the *line integral* of the magnetic force along that line, and we see that it is independent of the radius, and therefore of the form of the path. Accordingly, the line integral of the magnetic force along any closed line embracing a current is equal to 4π times the total current flowing through that closed line. Applying then Maxwell's principle, we see that if in any dielectric the electric

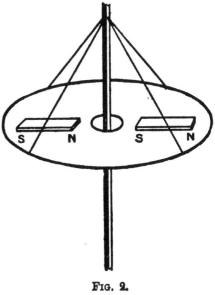

Fig. 2.

strain (D) is changing with time, its rate of change $\dfrac{dD}{dt}$, or $\dot{D}$ multiplied by 4π, gives us the line integral of the magnetic force due to it along a boundary-line perpendicular to the direction of the electric strain. When this boundary-line encloses a very small area, the quotient of the line integral by the area is called the *curl*, and we may, therefore, express the above statement in symbols as follows :—

$$4\pi \dot{D} = \text{curl of } H$$

In the next place, the electromotive force in any circuit is defined as the line integral of the electric force (E) along that circuit, and Faraday's law of induction, as extended by Maxwell to dielectrics, tells us that the time variation of the magnetic flux through any area is a measure of the electromotive force in, or line integral of electric force round, that bounding line.

Hence, when dealing with a unit area we can express the above fact symbolically, thus

$$-\dot{B} = \text{curl of } E$$

The minus sign is prefixed to $\dot{B}$ because a diminution of the flux is required to produce a right-handed or positively directed electric force.

The two equations

$$4\pi\dot{D} = \text{curl of H (magnetic force)}$$
$$-\dot{B} = \text{curl of E (electric force)}$$

establish a cross connection between the quantities D and B and E and H. There are also two direct relations, viz.

$$B = \mu H$$
$$4\pi D = KE$$

which express the fact that the magnetic flux B is proportional to the magnetic force H and to the magnetic permeability μ, and also that the electric strain D is proportional to the electric force E and to the dielectric constant K.

This last equation is obtained in the following manner. If we suppose a sphere of radius r described in a dielectric, and that a small conductor charged with Q electrostatic units of electricity is placed at the centre, then through the surface of the sphere there will be an electric strain D produced, which is everywhere directed outwards along the radius, and the product $4\pi r^2 D$, or the surface integral of the strain through the whole surface, will be equal to the quantity Q put at the centre.

Therefore $\qquad\qquad 4\pi r^2 D = Q$

or $\qquad\qquad\qquad 4\pi D = K\left(\dfrac{Q}{Kr^2}\right)$

But $\dfrac{Q}{Kr^2}$ is the radial electric force E at the surface of the sphere.

Hence $\qquad\qquad 4\pi D = KE$

Accordingly, from the above four expressions we can deduce the two important equations of electromagnetism, viz.

$$4\pi\dot{D} = K\dot{E} = \text{curl H}$$
$$-\dot{B} = \mu\dot{H} = \text{curl E}$$

The dot over a letter signifies the rate of change with time of the quantity denoted by that letter, and the expression " curl " signifies the line integral round a unit of area, or the quotient of the line integral round any small area by the magnitude of that area. If we then translate these conceptions into common language we can describe the production of an electromagnetic or electric wave in a dielectric as follows :—

To produce an electric wave we must first create at some place in the dielectric or in the æther a very sudden change in an electric strain. This may, for instance, be the sudden release or destruction of a very intense localised electric strain. The results of this, in accordance with Maxwell's first principle, is to generate magnetic flux along a circular line embracing the decreasing strain, the said flux line having its plane perpendicular to the direction of the strain. This initial strain is represented as to direction by the arrow in Fig. 3, and the embracing circular line of flux by the

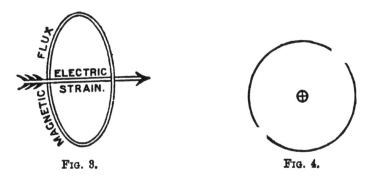

FIG. 3.   FIG. 4.

double elliptical line, being a circle seen in perspective. We may also represent the end-on view of the strain by a small circle, containing a cross as in Fig. 4, and the embracing flux line is then represented by the larger embracing circle. When the end-on view of a line of electric strain or magnetic flux is represented by a small circle, we may represent the direction of the strain or flux, whether to or from the reader, by placing a *dot* or a *cross* in that circle, the dot representing that the strain or flux is towards the reader, and the cross that it is away from him.

The reader should also bear in mind that a diminishing electric strain is by Maxwell's first principle equivalent to a current in an opposite direction to the strain, and an increasing strain to a current in the same direction as the strain. Also that the relation between the direction of a current and of the direction of its circular embracing magnetic flux is that of the thrust and twist of a corkscrew.

Again, by Maxwell's second principle, the creation of a line of magnetic flux, or its strengthening, involves the production of electric strain along lines embracing the flux line, so related as to direction that on any section plane transverse to the flux line the

electric strain is counter-clockwise in direction round that part of
the section in which the flux is away from the reader, and clockwise around that section of the flux which is towards the reader.
Bearing this in mind, it will be seen that the diminution of the
original central electric strain is accompanied by the production of
a series of concentric circular lines of magnetic flux all embracing
the original line of strain, and with directions as denoted in Fig. 5.

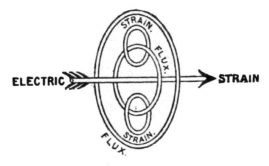

Fig. 5.

These flux lines are not, however, created simultaneously. The
flux lines nearest the original strain are generated and increase as
the original strain dies away, and this creation of magnetic flux
involves the production of other closed loops of electric strain
linked with these flux lines, which new strain lines tend, as they
are produced, to create in turn other circular lines of magnetic
flux in the same direction as the first created, but lying further
away from the line of original strain. In this manner a state of
alternate electric strain and magnetic flux is created at continually
increasing distances, and is propagated out into the medium. At
the instant when the original central electric strain has disappeared
the line of flux nearest to it, due to its variation, has disappeared
also, but the disappearance of this flux involves the creation of
lines of electric strain linked with it, which in the interior of the
circle of flux have a direction opposite to the original strain and
outside of it the same direction. This is equivalent to a reversal
of direction of the inner original or central strain, and it therefore
involves a reversal in direction of the whole system of embracing
and co-linked lines of flux and strain. This reversal in direction
does not, however, take place simultaneously at all points of space,
but successively from point to point outwards into space. A
careful consideration of the diagrams will therefore show that as
the original energy of electric strain imparted to the medium
disappears, it expends itself in creating an equivalent in the form

of an embracing magnetic flux, and that this flux in turn expends its energy in creating electric strain in the medium outside its line. Hence the energy imparted at one point is transferred from point to point in the medium by an action between contiguous parts of it, the principle involved being that variation or change of electric strain produces embracing magnetic flux, and variation or change of magnetic flux produces electric strain. The directions of the strain and flux are at right angles to each other, whilst both take place along lines which are self-closed or circuital. The portion of the medium which is the seat of these actions is continually changed; in other words, the operation is propagated through the medium with a velocity which is definitely measurable and related to the specific qualities of it. Hence the sudden release of an electric strain at one place is felt after a time at regions far removed from it. If this phenomenon is considered it will be recognised to be exactly analogous to the effect produced upon the surface of still water when we make a sudden depression at one point, as by a throwing a stone into it. We have then a depression of water *level* at the point of impact succeeded by elevation, and we have water *motion* set up as the result of these changes of level, when the water moves up or down to remove them. The changes of level correspond to the electric strain, and the water motion to the magnetic flux. Change of level of water results in the production of motion in the water, and motion in virtue of inertia results in the production of change of level, just as in the dielectric, change of electric strain results in the production of magnetic flux, and change in flux produces electric strain.

If we take a section of the electric field transversely to the original flux, we find that the field is occupied with concentric lines of magnetic flux, and orthogonally, or at right angles to these, and intermixed are packed the cross-sections of lines of electric strain, the strain being orthogonal or at right angles to the flux. Likewise, if we take a section of the field in the plane of the original electric strain, we find it to be occupied by closed loops of electric strain, intermixed with which are found the cross-section of lines of magnetic flux.

Moreover, the direction in which the transmission of the energy is taking place is at right angles to the plane which contains the directions of the lines of magnetic flux and electric strain.

Thus, an electromagnetic wave is, so to speak, woven out of electric strain and magnetic flux which constitute respectively the warp and the weft of the fabric. The magnetic flux and electric strain are called the component magnetic and electric vectors which make up the wave.

The magnetic component at any one spot changes cyclically in strength and reverses or alternates in direction, although it may maintain a constant direction. So also does the electric component, but at the same point in space the electric component is a maximum at the instant when the magnetic component is zero, and *vice versâ;* in other words, the two vectors differ 90° in phase.

We can imagine ourselves endowed with special senses of such a kind as to enable us to detect the presence of lines of magnetic flux and lines of electric strain in space, and appreciate their direction and movement. Suppose, then, that we took up our position at a fixed spot in space through which electromagnetic waves were passing. We should detect these regions of magnetic flux and electric strain, alternately succeeding one another at that place. On the other hand, we can imagine it possible for us to fasten our attention upon a particular phase of the strain or flux, and to move along so as to keep always in contact with that same phase of the force or flux. We should then find ourselves travelling in a certain direction normal to the direction of the flux and strain, with a velocity called the wave velocity.

5. **The Velocity of an Electromagnetic Wave.**—Let us next consider the propagation of a plane electromagnetic wave in which the electric and magnetic components are at right angles to each other and to the direction of propagation. Suppose the electric force (E) is everywhere parallel to the axis of x, the magnetic force (H) to the axis of y, and the direction of propagation of the wave to the axis of z (see Fig. 6).

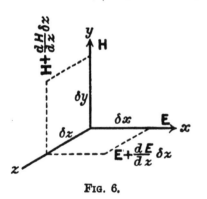

FIG. 6.

The student who has even a small knowledge of the principles of the differential calculus knows that if the ordinate of a curve corresponding to any abscissa x is denoted by y, then the ordinate corresponding to an abscissa $x + \delta x$, where δx is a small increase in x, is $y + \frac{dy}{dx}\delta x$. Hence, if at the origin the electric force has a value E along the axis x, then the electric force in the plane of xz in a direction parallel to the axis of x, but at a distance δz from it, is $E + \frac{dE}{dz}\delta z$, being greater the further we remove from the origin.

If, then, we take the line-integral of the electric force round a small rectangle, whose sides have lengths δx, δz respectively, lying in the plane of xz, with one corner on the origin, we have to sum up all round this rectangle the product of the length of each side by the electric force along that side, travelling always in the same sense round the rectangle, and reckoning the product as positive when the force is in the same direction as the motion, and negative when it is against it. We note, then, that parallel to the two sides of length δx, the force has a value E and $E + \dfrac{dz}{dE}\delta z$ respectively, and that parallel to the two sides the electric force is zero, because there is no component in the direction of propagation. Hence the line integral is equal to

$$\left(E + \frac{dE}{dz}\delta z\right)\delta x - E\delta x = \frac{dE}{dz}\delta z\delta x$$

If we divide this line integral by the area $\delta z \delta x$ of the rectangle, we have the curl of the electric force in the plane of xz, viz. $\dfrac{dE}{dz}$.

We have, however, shown in the last section that the curl of the electric force is equal to the time rate of decrease of the magnetic flux through the area or to $-\dot{B}$ or to $-\mu\dfrac{dH}{dt}$. Hence we have the equation

$$-\mu\frac{dH}{dt} = \frac{dE}{dz}$$

as one relation between the electric and magnetic forces. We can then obtain another by taking the curl of the magnetic force in the plane yz, for it is obvious that if the magnetic force is H at the origin, then it is $H + \dfrac{dH}{dz}\delta z$ in the plane of yz at a distance δz from the origin parallel to the axis y. Hence the line integral round a rectangle $\delta y \cdot \delta z$ with its corner on the origin is

$$H\delta y - \left(H + \frac{dH}{dz}\delta z\right)\delta y = -\frac{dH}{dz}\delta z\delta y$$

and therefore the curl of H in the plane yz is $-\dfrac{dH}{dz}$.

We have already shown that the curl of the magnetic force is equal to $K\dot{E}$ or to $K\dfrac{dE}{dt}$.

Hence we have a second relation between the electric and magnetic forces, viz.

$$K\frac{dE}{dt} = -\frac{dH}{dz}$$

Combining this with the previous one, it is easy by differentiation to separate the variables, and to obtain the two simultaneous equations,

$$\frac{d^2H}{dt^2} = \frac{1}{\mu K}\frac{d^2H}{dz^2} \quad \cdots \cdots \quad (1)$$

$$\frac{d^2E}{dt^2} = \frac{1}{\mu K}\frac{d^2E}{dz^2} \quad \cdots \cdots \quad (2)$$

The above equations are of a type which constantly occurs in various branches of physics, for instance in acoustics, and they have various solutions applicable to different problems. The reader, however, will find by trial that they can be satisfied by solutions of the following form

$$E = E_0 \sin 2\pi\left(\frac{z}{\lambda} - \frac{t}{T}\right) \quad \cdots \cdots \quad (3)$$

$$H = H_0 \sin 2\pi\left(\frac{z}{\lambda} - \frac{t}{T}\right) \quad \cdots \cdots \quad (4)$$

provided that

$$\frac{T}{\lambda} = \sqrt{\mu K}$$

The student should differentiate each of the above equations (3) and (4) twice with regard to t, and twice with regard to z, and then prove that consistently with the last-named relation they satisfy the above differential equations (1) and (2) of the second order. These solutions imply that E and H are quantities which are periodic in space and also in time. For if we consider z to be a constant quantity, then the result of giving to t gradually increasing values in the expressions for E and H, results in periodic values for these quantities, both E and H being in step with each or being zero and a maximum at the same instant. If we keep t constant and plot out each of the above equations for E and H in terms of z, we find we obtain a wavy line or sine curve in each case. Hence these expressions for E and H represent waves.

The reader should then take two slips of card and slit them both halfway down the middle, and then set them one in the other so that they form two planes intersecting at right angles (see Fig. 7). On each card should then be drawn a wavy line, the lines being so set that the maximum and zero points coincide, as shown in Fig. 7. One of these lines should be marked E, and the other H. The ordinates of one curve will represent the

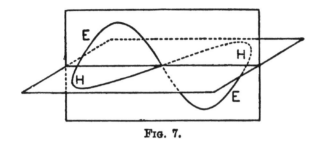

FIG. 7.

variation of the magnetic force as we proceed along the line of intersection of the planes, and the ordinates of the other the variation of the electric force. Moreover, the equations above (3) and (4) show us that E and H have the same values if z and t both increase at the same time, so that when z becomes $z + z'$, t becomes $t + t'$, provided that z' and t' are so related that $\dfrac{z'}{t'} = \dfrac{\lambda}{T}$; in other words, if $\dfrac{z'}{t'}$ is a velocity equal to the velocity of the wave, viz. $\dfrac{\lambda}{T}$.

But now $\dfrac{T}{\lambda}$ is equal to the quantity $\sqrt{\mu K}$, and accordingly it follows that the ratio $\dfrac{1}{\sqrt{\mu K}}$ must be the velocity of the electro-magnetic wave.

The next question is to determine numerically the above velocity. We do not know the absolute values of the permeability and dielectric constant of any medium or dielectric. We are in the habit of assuming for the purposes of comparison that these quantities are unity for air or æther. There are two systems of measurements and two sets of units, in one of which we assume $K = 1$ for air or æther, and in the other that $\mu = 1$, and these are called the electrostatic and electromagnetic systems of unity respectively. Thus, for instance, we agree that the electrostatic unit of electric quantity shall be such a quantity that when placed upon a small sphere it will repel another equal charge on a similar sphere at a distance of 1 centimetre with a force of 1 dyne or $\frac{1}{981}$ part of the weight of 1 gramme, provided the experiment is made in air or vacuum. If, however, we were to make the measurement with the balls immersed in paraffin oil, the absolute quantity required to create this same repulsive force would be about $1\cdot4$ times or $\sqrt{2}$ times greater, because the force varies as the square root of the dielectric constant of the surrounding medium, which in this case is nearly twice that of air. Hence the absolute value of the electrostatic unit of quantity varies as $\sqrt{\overline{K}}$.

In the same manner we define the electrostatic unit of current as the current which conveys 1 unit of quantity per second. Hence, this also varies as $\sqrt{\overline{K}}$, where K is the dielectric constant of the medium. We may, however, measure a current by the magnetic force it produces, or the mechanical force on a unit magnetic pole placed at a unit distance from a unit length of the current, and this in turn depends upon the magnitude of the unit magnetic pole.

If two conductors carrying equal currents attract or repel each other with a certain force in air, then, if immersed in a medium of greater permeability, the current would have to be decreased to maintain the same force between them. The unit of current, therefore, measured magnetically varies inversely as the square root of the permeability of the medium, or as $\dfrac{1}{\sqrt{\mu}}$. Hence the ratio of the electrostatic unit of current to the electromagnetic unit of current is the ratio of $\sqrt{\overline{K}}$ to $\dfrac{1}{\sqrt{\mu}}$ or the ratio of $\sqrt{\mu K}$ to 1.

By certain purely electrical measurements which are described in treatises on electricity, this ratio of the units can be determined, and has been determined many times with great care. It is found to be a ratio not far from 1 to 3×10^{10}. In other words, for air or aether, the value of the quantity $\dfrac{1}{\sqrt{\mu K}}$ is numerically 3×10^{10} when using the centimetre, gramme, and second, as our units of length, mass, and time, or 300,000 if we use the kilometre, gramme, and second.

Hence, although we do not know the absolute values of the dielectric constant and magnetic permeability of the electro-magnetic medium, we do know that the reciprocal of the square root of their product is a number identical with that of the velocity of light when measured in the same units. But we have proved that this quantity is the expression for the velocity of an electro-magnetic wave through space. Hence the conclusion is that the velocity of an electromagnetic wave through space is the same as that of light, and the conviction is immensely strengthened that the medium concerned in the two phenomena must be one and the same.

We give below a few of the recent determinations of the velocity of light and of the electromagnetic velocity determined by means of the ratio of various electric and magnetic units.

Experimental Determinations of the Velocity of Light.

Observer.	Velocity in kilometres per second.
Cornu	300,040
Michelson	299,853
Newcomb	299,860
Perrotin	299,860

The most probable value according to Weinberg is

$$299,852 \frac{km}{sec} \text{ or } 2.99852 \times 10^{10} \text{ centimetres per second.}$$

Experimental Determinations of the Ratio of the Electrical Units or of the Electromagnetic Velocity $= \dfrac{1}{\sqrt{K\mu}}.$

Observer.	Value in kilometres per second.
Himstedt	300,570
Rosa	300,000
Thomson and Searle	299,600
H. Abraham	299,130
Pellat	300,920
Hurmuzescu	300,100
Perot and Fabry	299,780

The mean value of $\dfrac{1}{\sqrt{K\mu}}$ is $300,040 \dfrac{km}{sec}$ or 3×10^{10} centimetres per second.

Hence the electromagnetic velocity is in close agreement with the velocity of light.

6. The Practical Production of Electromagnetic Waves.—The principles discussed in the previous sections of this chapter were all fairly well understood prior to 1887, but yet no one had been able to put them into practice in such a manner as to generate electromagnetic waves at pleasure and detect them when made.

G. F. Fitzgerald had, however, suggested that they might be produced by the high frequency oscillations of a Leyden jar. Lodge had studied the phenomenon in great detail, and had shown how to create stationary electric oscillations on wire with loops and nodes of potential. It remained, however, for Heinrich Hertz, in

1888, by a stroke of genius to place the world in possession of the secret by his invention of the open radiative oscillatory circuit, and what was equally important by the discovery of a simple means for detecting and measuring the wave length of the electromagnetic radiation. When this solution of the problem was given it became at once obvious that this method for the production of electromagnetic waves consisted in establishing an electric strain at some place which is localised or chiefly confined to a small region, and is then very suddenly released, and this starts into existence the train of operations in the dielectric resulting in an electromagnetic wave.

For an historical account of the discovery, the student must be referred to larger treatises; we shall here confine ourselves mainly to an elucidation of principles. Suppose, then, that two long metal rods are furnished with rounded ends or spark balls at one end, and are placed in one line with the balls in close approximation and the rods both insulated (see Fig. 8). These two rods constitute the

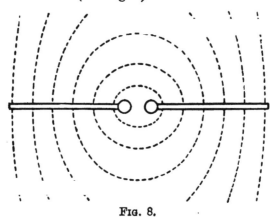

FIG. 8.

metallic surfaces of a condenser of which the dielectric is the surrounding air and æther. If, then, we connect them to any source of high tension electric supply, such as the terminals of the secondary circuit of an induction coil, we can charge these rods with positive and negative electricity and bring them to a large difference of potential. When this is the case lines of electric strain stretch from one rod to the other along curvilinear paths, as shown in Fig. 8 by the dotted lines, and between the contiguous ends of the rods in the small air gap the electric field is very intense. When this field reaches a certain intensity the air in the gap, which up to a certain strain is an excellent insulator, passes quite suddenly into a highly conductive condition, and the result is that the two oppositely charged rods forming the two surfaces of a condenser are connected through a low resistance. The charges then oscillate in the manner described in Chapter I. in discussing the oscillatory discharge of a Leyden jar. The capacity of the rods relatively to each other being extremely small and the inductance of the discharge path small also, the frequency of these oscillations is very great, even as high as several million per second, although owing to the radiative power of the system the oscillations in a train are very few, being damped out almost immediately.

Nevertheless, we have here all the conditions necessary for the production of an electromagnetic wave. We have in the small air space between the ends of the rods an intense electric strain produced by the opposite charges of the rods, and this strain is almost instantly abolished when the air in the spark gap passes into the conductive condition. An electric current is then created across the gap, and the potential difference of the rods rapidly falls. This current, however, can only take place in a circuit, and the circuit here involved is composed partly of a metallic part forming the rods and partly of the dielectric outside. The current in the rods is called a current of conduction, and the current in the dielectric a displacement current or change of electric strain. Hence, as the original dielectric strain dies away it expends its energy in producing a conduction current in the rod, and this last in turn a gradually increasing strain in the dielectric outside, which is in the reverse direction to that creating the conduction current. In virtue of inductance or electric inertia this conduction current tends to persist, and finally expends its energy in re-creating the dielectric strain in an opposite direction to that in which it was originated. In other words, the electric charges of the rods are then also reversed, since the direction of the electric strain in the space round them is reversed. The energy originally imparted to the rods therefore changes rapidly from an electrostatic or potential form to an electromagnetic or kinetic form, and the rods are alternately at large differences of potential with no current passing the centre or air gap, and then at nearly the same potential, but with a large current passing the centre and expending itself in again charging up the rods to a high potential difference so as to repeat the process until the energy is dispersed. In consequence of these sudden variations in the electric strain between the adjacent ends of the rods, they create electromagnetic radiation or damped electric waves in the manner already described until the originally stored energy has all been radiated. As often as the oscillations in a train die away, the air in the air gap comes back to its original non-conductive condition so that the operation can be repeated, and if the source of supply is an induction coil actuated with an automatic break, we have a continually repeated oscillatory discharge accompanied by a sharp crackling spark in the air gap between the balls, provided the spark balls are set at the right distance. As a consequence we have the continual radiation from the rods of trains of highly damped electromagnetic waves, the trains succeeding each other at a rate determined by the induction coil break, but each train consisting probably of not more than four or five oscillations. As regards the disposition of magnetic and electric force in the space around the rods, it is easy to see that the magnetic force will be distributed in circles arranged with their centres on the axis of the rods, since the current takes place backwards and forwards along the rods (see Fig. 9).

As regards the electric force the distribution and form of the strain lines is somewhat more complex. If we de-

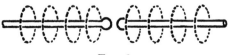

FIG. 9.

scribe round the rods a sphere having a radius equal to the length of either rod, it can be shown that outside this sphere the electric force is distributed in certain closed curves or loops which are concave on the side facing the rod, and that from instant to instant these loops change their position during the phase in a manner which is equivalent to a movement of each loop outwards from the rods in its own plane and an expansion of the loop at the same time, as shown in the series of diagrams in Fig. 10, delineated by Professor K. Pearson and Miss Alice Lee. If, however, we confine attention to the neighbourhood of a plane drawn through the spark gap, and at right angles to the rods, called the equatorial plane, it is obvious that the magnetic flux is arranged at any instant in concentric circles in this plane, and that the electric strain is everywhere perpendicular to it and to the magnetic flux, whilst the direction of propagation of the wave is radially outwards along this plane in all directions equally (see Fig. 10). The predetermination of the direction of the electric force in other regions round the rods and particularly near the ends of the rods is an investigation which is more difficult, and cannot be conducted without mathematical analysis for which the student must be referred to more advanced text books.[1]

[1] For the full explanation of this rather complicated phenomenon the reader must be referred to the discussion of it in Chapter V. of the Author's book "The Principles of Electric Wave Telegraphy." (Longmans, Green, and Co.)

An arrangement of two rods, as above described, with a spark gap in the centre constitutes the simplest form of *linear radiator* or *Hertzian Oscillator* for the production of damped electro-

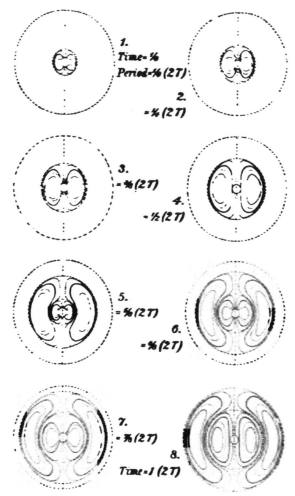

FIG. 10.—Delineation of the lines of electric strain round a small oscillator taken at equal intervals during one complete oscillation. Drawn by Professor Karl Pearson and Miss Alice Lee.

magnetic waves. It can be shown that the wave length of the emitted radiation is about 2·5 times the total length of the two rods, and that the energy (*e*) in ergs emitted in the form of electromagnetic waves outwards through the surface of a sphere described round the rods, of radius large compared with their length, per complete period of the oscillation is given by the expression—

$$e = \frac{16\pi^4\phi^2}{3\lambda^3}$$

where ϕ denotes the *electric moment* of the oscillator.

This last term is defined as follows: Imagine an ideal oscillator consisting of two small spheres connected by a thin rod. We may consider that the rod possesses inductance but negligible capacity, and the whole of the capacity is in the sphere. During the oscillations a certain quantity of electricity may be considered to oscillate backwards and forwards, and the product of this quantity by the distance between the spheres is the electric moment. In an actual linear oscillator the effective length is something less than the real length, just as in the case of the moment of a magnet, the actual distance between the poles, which is one of the factors of the magnet moment, is indeterminate, but something less than the real length. The electric moment of an oscillator can, however, be measured as a single quantity, like the moment of a magnet, and is proportional to the original charge given to it, and therefore to the capacity of one-half of the oscillator with reference to the other and to the initial potential difference to which they are charged. It is also proportional to the length of the oscillator. Since, then, the wave length of the emitted radiation is proportional to the length of the oscillator, we see that the above formula for the radiation per period of the linear oscillator shows us that this radiation will be proportional to the square of the capacity, to the square of the charging voltage, and inversely as the wave length or directly as the frequency of the oscillations.

In his historical researches Hertz constructed his oscillator by attaching to one end of two short, rather stout rods of brass, two square zinc plates, and furnishing the other ends of the rods with spark balls. These rods were placed in line with spark balls in apposition, and when so arranged the plates with the surrounding air as dielectric formed a condenser of a certain capacity. When plates are charged by connecting them to the secondary terminals of an induction coil, and the spark balls on the rods approached to within 4 or 5 millimetres of each other, a bright crackling spark passes and oscillatory discharges take place across the gap. The result of these oscillations is to create, as we have seen, electric radiation, and the oscillations expend their energy in creating the state of alternate electric strain and magnetic flux which constitutes an electric wave. The characteristic of a wave is that energy is conveyed away from the wave-making body and exists in the surrounding medium in a double form partly potential and partly kinetic, the energy of the complete wave being half potential and half kinetic, and transferred from point to point in the medium with the velocity of light.

We must consider a little more closely the operations which are taking place. When the two rods or parts of the oscillator are charged to different potentials, energy is stored up in them. If C is the capacity of one-half of the oscillator with respect to the other in microfarads, and V is the potential difference in volts, then $\frac{1}{2}\frac{C}{10^6}V^2$ is the energy storage in joules and $\frac{10}{2}CV^2$ is the storage reckoned in ergs. When the discharge takes place a certain proportion of this energy is dissipated as heat and light in the spark, and also a small proportion as heat produced in the the rods by the oscillatory current taking place in them. The larger portion, however, is communicated to the dielectric in the form of magnetic flux and electric strain, and this part does not return in again upon the oscillator, but is permanently imparted to the dielectric as an electromagnetic wave which travels away from the oscillator. The damping of the oscillations is thus partly due to resistance and partly to radiation, and the total logarithmic decrement is made up of two parts, viz. the *resistance decrement* and the *radiation decrement*, and accordingly as this last coefficient is large or small, so is the oscillator called a good or a poor radiator. The laws which govern electric radiation are closely analogous to those of radiant heat and light, and this is what we might expect, seeing that in both cases we are concerned with the vibrations of the same medium—the æther—although with different wave lengths.

CHAPTER V

RADIATING AND RECEIVING CIRCUITS

1. Varieties of Radiative and Receiving Circuits.—Broadly speaking, circuits in which we can establish oscillations may be divided into *open* and *closed* circuits which correspond respectively to good and bad radiators of electromagnetic waves. If we place two rods or wires, each ·having at one end a spark ball, and at the other end a metal plate, in one line with the two spark balls in close proximity, and the plates as far apart as possible (see Fig. 1 (*a*)),

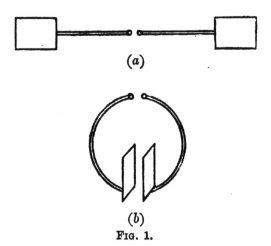

(*a*)

(*b*)

Fig. 1.

we then construct a circuit called an open radiative circuit which is a very good radiator or producer of electromagnetic waves. If, however, we bend round the wires, so that the plates come into approximation (see Fig. 1 (*b*)), we increase their electric capacity with respect to each other and form a closed radiative circuit which is relatively to the open one a poor radiator. Between these two extremes there is, however, no hard and sharp separation, and we may have every possible intermediate variety of radiative circuit. These various electric circuits correspond as regards their electric radiative power to good and bad thermal radiators. A rough black surface is a good radiator of heat, and a bright polished silver surface a poor one. Moreover, it is a fundamental principle in thermal radiation that good radiators are good absorbers. Thus a lamp-blacked surface is not only a good radiator of heat, but readily absorbs heat falling upon it. In the same way good electromagnetic radiators are good absorbers of electromagnetic waves. If such waves travelling through space fall in the right direction upon an open or closed oscillatory circuit, that is, one having capacity and inductance in series, the energy of the waves is absorbed to a greater or less extent, and expends itself in setting up electric oscillations in the receiving circuit.

A precise statement was long ago formulated as regards radiant Heat and Light called the Law of Exchanges, which is the foundation of spectrum analysis, viz. that a body absorbs best those particular luminous or thermal radiations which it emits if heated. An identical law holds good in the case of electric radiation, which may be enunciated as follows :—

An oscillatory circuit, or one having capacity and inductance, in which therefore electric oscillations can be excited, absorbs some of the energy of electromagnetic radiation falling upon it, and it absorbs best radiation of that kind and wave length which it would itself emit if set in oscillation, and absorbs it most readily when arriving in the direction in which it would itself radiate most strongly.

This law of electromagnetic radiation is of the greatest importance, and may be said to be the foundation on which the art of radiotelegraphy is erected.

In the case of an open circuit electric radiator the free ends become at intervals the seat of electric charges which create an electric potential in space, and as the electric strain at any point in the surrounding space is partly due to these free charges, an open radiative circuit is also called an *electric oscillator*. In the case of a closed radiative circuit the plates being near together and carrying electric charges of opposite sign, these tend to neutralise each other's effect in external space, and the predominant agency in creating the radiation is therefore the current in the circuit. Hence, a closed radiative circuit is sometimes called a *magnetic oscillator*.

A large variety of open and closed or intermediate forms of electric and magnetic oscillator are employed in radiotelegraphy.

2. **The Open Circuit Oscillator.**—This radiator consists of a vertical or nearly vertical rod or wire A, the upper end of which is insulated, and may or may not terminate in a metal plate, whilst the lower end of the rod or wire is connected to a good conducting plate (E) buried in the earth, or placed near the surface of the earth (see Fig. 2). It is usually called an *aerial, air-wire,* or *antenna*. The simplest method of establishing oscillations in this antenna consists in interrupting the wire just above the earth and interposing a pair of spark balls. These balls are then connected

to the secondary terminals of an induction coil, and when the coil is in action the upper part of the aerial is charged at intervals, say, with negative electricity. When this charge reaches a certain potential the insulation of the air in the gap breaks down and electric oscillations are thus set up in the antenna. Prior to the discharge the wire itself forms one plate of a condenser, of which the surrounding earth is the other coating and the air and æther the dielectric. In this condition lines of electric strain stretch from the wire to the earth on all sides (see Fig. 2). When the discharge takes place this condenser is discharged and the electric strain in the spark gap disappears, owing to the air in the gap becoming conductive. At this moment the charge in the antenna rushes down into the earth, creating a *conduction current* in the

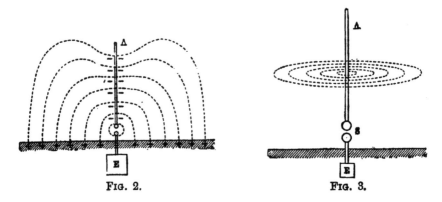

FIG. 2.   FIG. 3.

wire, which varies from point to point, but is a maximum at the earthed end. This conduction current is completed and made circuital by the simultaneous production of a *dielectric current* by the release of the dielectric strain in the space round the antenna. The antenna is thus surrounded by a magnetic field, the lines of flux being circles with their centres on the antenna (see Fig. 3). As the current in the wire continues, it builds up a new electric strain around the antenna, the direction of which is opposite to that strain, the relaxation of which produced the conduction current, and finally the kinetic energy of the conduction current is transformed entirely again into potential energy of electric strain outside the antenna. This process is continually repeated, and we have then an oscillatory electric current in the wire and periodic changes in the direction of the magnetic flux and electric strain at all points outside the wire, which are propagated outwards with the velocity of light. If we follow out in detail, by mathematical analysis, the movements of the lines of electric strain which originally stretched from the antenna to the earth, we find they are continually displaced, and the result of the first oscillation is that the ends of these lines which originally terminated on the antenna run down it, and finally so place themselves that they form semi-loops of electric strain with their ends resting on the earth (see Fig. 4). This detachment of the line of electric strain

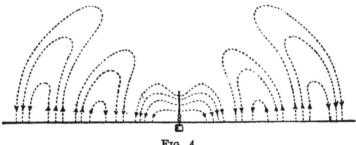

FIG. 4.

corresponds to and expresses the detachment of the energy which takes place from the antenna in the radiation, and at each oscillation a fresh production of such semi-loops of strain then takes place, those first produced moving away from the antenna radially in all directions. For the purposes of radiotelegraphy we may fix our attention entirely on the regions near the earth and at some distance from the antenna. In this district the magnetic flux lines are parallel to the earth, provided the antenna is vertical or nearly so, and the electric strain lines are perpendicular to the earth. At any one instant the electric strain at the earth's surface is directed alternately upwards and downwards over annular regions or districts which succeed each other radially. At each oscillation these regions are displaced outwards from the wire in every direction. The shortest distance between two adjacent places at which the electric strain has its maximum value in the same direction at the same time, is called a *wave length* for that antenna. For a simple or plain single wire antenna earthed at the m with a spark gap placed near the earthed end, the wave length is not far from 4·8, or, say, 5 times the length of the wire. A plain aerial wire of this kind has a relatively small capacity. If the wire is a circular-sectioned metal wire 0·1 inch or 2·5 mm. in diameter (*d*), and 100 feet or 300 cms. in length (*l*), its capacity in space far removed from the earth can be approximately calculated by the formula in Chapter I., viz.

$$\text{Capacity in microfarads} = \frac{l}{4\,6052 \times 9 \times 10^5 \times \log_{10}\dfrac{2l}{d}}$$

In this case $\dfrac{2l}{d}$ is 24000, and $\log_{10} 24000 = 4\cdot3802$, so that the capacity is $\dfrac{1}{3433}$ of a microfarad. The actual capacity of such a vertical insulated wire, with its lower end near the earth, would, however, be about 10 per cent. greater than the value calculated by the above formula, owing to the proximity of the earth.

Hence, even if charged to 30,000 volts, which is the equivalent of a 1 centimetre spark at the spark balls, the actual quantity of electricity put into such a single wire antenna would be only 6 microcoulombs, and the energy stored before discharge only about $\dfrac{1}{12}$ of a joule, or 83,000 ergs, for the energy put in is measured by the value of the expression $\dfrac{CV}{2}$ reckoned in consistent units.

The natural time period of oscillation, T, of this plain ærial is equal to $\dfrac{\lambda}{3 \times 10^{10}}$, where λ is the wave length of the radiation in centimetres, or to $\dfrac{4\cdot8l}{3 \times 10^{10}} = \dfrac{16l}{10^{11}}$ seconds, where l is the length of the wire in centimetres. Thus, for a single wire ærial 100 feet long it is $0\cdot48$ microsecond, which means that each complete oscillation takes place in rather less than one-half of a millionth of a second. Such a simple straight wire has, however, enormous radiative power. It can be shown from theoretical considerations that the *radiation decrement* (δ) per half period of a plain aerial wire of length l centimetre and capacity C microfarads is very nearly given by the formula

$$\delta = 2\cdot0 \times 9 \times 10^5 \frac{C}{l} = 1\cdot8 \times 10^6 \frac{C}{l}$$

It is therefore equal to nearly twice the capacity per centimetre of length reckoned in micro-microfarads.

Thus, if $l = 100$ feet $= 3000$ cms., and $C = \dfrac{1}{5000}$ microfarad, we have $\delta = 0\cdot13$, which implies that the Napierian logarithm of the ratio of one oscillation to the next in the opposite direction is $0\cdot13\cdot$

The ratio of one oscillation to the next in the opposite direction is $\epsilon^{-\delta}$, where $\epsilon = 2\cdot718$, viz. the base of the Napierian logarithms, and δ is the decrement per semi-period. Hence, in 10 complete oscillations, or in 20 semi-oscillations, the amplitude is reduced to a fraction of the initial amplitude, represented by $\epsilon^{-20\delta}$, which for the case in question is equal to $(2\cdot718)^{-3} = 0\cdot05$ nearly, or 5 per cent. of the initial value.

So that in less than a dozen periods the oscillations are practically damped out. The above formula, however, gives us only the decrement for radiation. In addition, there is some damping due to the resistance of the aerial wire itself and to the spark, so that in practice the actual number of oscillations produced when a plain aerial is charged and discharged is something less than half a dozen. This antenna is therefore called a *highly damped radiator*. The great radiative power of open oscillators makes it necessary to supply them with very large amounts of power to maintain in them persistent oscillations of high frequency. Thus, if we consider the case of a simple linear oscillator consisting of two rods in one line of total length l and capacity C microfarads with respect to each other, and if undamped oscillations of simple harmonic type and frequency, N, are maintained in this oscillator, the R.M.S. value (a) of the current at the centre, reckoned in amperes, would be given by the expression

$$a = \frac{2\pi N}{\sqrt{2}} \cdot \frac{C}{10^6} \cdot V$$

where V is the maximum potential difference of the rods in volts. The electric moment ϕ of the oscillator in electrostatic units would then be given by

$$\phi = \frac{9 \times 10^5 \, CVl}{300}$$

and the energy radiation (E) in ergs per period is

$$E = \frac{16\pi^4\phi^2}{3\lambda^3}$$

Substituting in this last expression the values of ϕ and CV from the two previous equations, we have for the value of the energy in ergs radiated per period the formula

$$E = 8\pi^2 \frac{l^2}{\lambda^2} \frac{a^2}{N} \times 10^8$$

and therefore the energy radiated per second, or the power (W) in watts supplied to maintain the oscillations continuously, is given by the equation

$$W = 80\pi^2 \frac{l^2}{\lambda^2} a^2$$

Now, for a simple double rod antenna, the ratio $\dfrac{l}{\lambda}$ is nearly 0 4, and, since $\pi^2 = 9\cdot87$, we have finally

$$W = 126a^2$$

Hence, to produce persistent oscillations with a current having an R.M.S. value of 10 amperes at the centre of the oscillator, would involve the expenditure of 12 kilowatts.

Bearing in mind that $N\lambda = 3 \times 10^{10}$, we can also write the value of the power absorbed in the form

$$W = 87 \times 10^{-20}\, l^2 a^2 N^2$$

which shows us that for the linear open oscillator the power absorbed to produce persistent oscillations varies as the square of the frequency, and as the square of the current at its centre.

3. **The Closed Circuit Oscillator.**—A closed circuit oscillator is made by constructing some form of loop of wire which may have its plane vertical or horizontal, and inserting in the circuit a suitable form of condenser.

It is quite easy to construct a horizontal closed circuit radiator by driving into the ground four or more stakes or telegraph poles which carry insulators, and then straining round these a wire or wires in parallel, so as to make a large loop of one turn, the ends of which are brought into a house and connected in series with a condenser and to a pair of spark balls, if damped oscillations are to be set up, or to an electric arc or alternator if undamped oscillations are required. More often a closed radiator has its plane set vertically. It is then necessary to erect some form of mast or tower to carry it. The simplest form of closed circuit vertical radiator is made by attaching to the top of a mast or tower two wires of about equal length which are upheld by one or two insulators at the summit, or under some circumstances this point of the loop may be uninsulated. The wires must be of considerably greater length than the height of the mast. The lower ends are then brought into a signalling house and attached to fixed points. To some point or points at or below the middle of the wires guy lines are attached by insulators and strained tightly so as to stretch out the wires and form a lozenge or triangular-shaped closed antenna (see Fig. 5), which may have its plane set in any required direction. Otherwise, two masts or towers are erected in required positions, and between these from insulators one or more horizontal wires are suspended, which are continued downwards in a vertical direction, and the ends brought into a signalling house, the whole arrangement forming a square or rectangular vertical closed circuit, which may have its plane set in any required direction. Or a single mast or tower may be employed, having two sprits attached to it by means of which an antenna wire is upheld in the form of a vertical rectangle, the two ends being brought into a signalling house. This last arrangement possesses the advantage that by swinging round the sprits the plane of the loop can be altered so as to set in any required direction, as required in the case of directive radiotelegraphy. In any case, the closed loop may consist of either a single wire or a number of wires arranged in parallel.

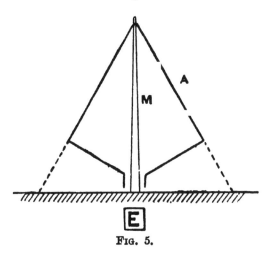

FIG. 5.

In the case of the closed circuit antenna the arrangement of a number of wires in parallel not too near together, has the advantage of reducing the inductance of the loop, and thus enabling a large area to be employed without correspondingly increasing the inductance. This same advantage is obtained by arranging wires in parallel in constructing an open circuit antenna, but in the case of the open antenna the objection exists that multiplying the wires increases the total capacity of the antenna although it decreases the inductance, so that one effect to a certain extent nullifies the other as regards the reduction of the time period of the antennæ.

The closed oscillating circuit as compared with the open has much less radiative power, and, therefore, a much smaller radiative decrement. If a closed oscillatory circuit of area S is traversed by a current of a maximum value I, then the product IS = M is called the maximum magnetic moment of the circuit, just as the product of the length of the open oscillator and the maximum electric charge at the extremity is called the maximum electric moment. It can be shown that the current radiated through a sphere of large radius described round a closed oscillator is given by the formula

$$E = \frac{16\pi^4 M^2}{3\lambda^3}$$

where λ is the wave length of the emitted radiation.

If the maximum current measured in amperes is denoted by A, and the area in square centimetres by S, then the magnetic moment of the closed oscillator M is given by $\dfrac{AS}{10}$. Hence, if a is the R.M.S. value of the current, we have

$$E = 10\cdot4\frac{Sa^2}{\lambda^3}$$

Accordingly, the power absorbed by the oscillator to radiate persistent undamped radiation in watts is given by

$$W = 4 \times 10^{-38} \times S^2a^2N^4$$

where N is the frequency, and $N\lambda = 3 \times 10^{10}$.

If we compare the above expressions for the power radiated by the closed and open oscillators, assuming them to be in both cases the seat of persistent oscillations, we see that we have the two expressions

$$W = 4 \times 10^{-38}\,S^2a^2N^4 \text{ (closed oscillator)}$$
$$W = 87 \times 10^{-20}\,l^2a^2N^2 \text{ (open oscillator)}$$

for the power absorbed in persistent radiation in the two oscillators respectively. These last two formulæ show us that the power radiated varies as the square of the frequency in the case of the open oscillator, but as the fourth power for the closed oscillator. Hence the power radiated by a closed oscillator increases very much faster with the frequency than in the case of the open one, and conversely for the closed oscillator it is very much smaller for the same frequency, and decreases very much more rapidly as the frequency is lowered. Accordingly, a closed oscillatory circuit has sometimes been called non-radiative, but in truth there is no such thing as an absolutely non-radiative circuit; it is a question of degree, and as in the case of thermal radiation some surfaces are better radiators of heat than others, but no surface is absolutely non-radiative, so in the case of electrical radiation, some circuits are vastly better radiators than others. The closed circuit oscillator has, however, certain valuable qualities as a directive radiator ; that is to say, its radiation is not equal in all directions round its vertical axis. In the case of a linear open oscillator it is obvious that since everything is symmetrical round the axis of the oscillator, the radiation must be equal in all directions in planes passing through this axis of symmetry, and the magnetic field of the open oscillator is, as already shown, distributed in circular lines with their centres on the axis. If, however, we consider the magnetic field of a closed circuit, say, in the form of a circle, and consider the distribution of the magnetic field in a plane perpendicular to the plane of the circuit drawn through its centre, it would be seen to be as shown by the dotted lines in Fig. 6, where the large black dots represent the cross-section of the circular circuit.

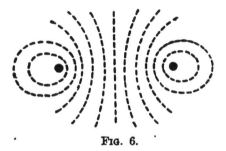

FIG. 6.

This matter will be dealt with more fully in a subsequent section; meanwhile it may be pointed out that the radiation of a closed circuit antenna is at a maximum in the plane of the circuit, and although symmetrical with respect to one plane is not symmetrical with respect to any one line drawn in it.

4. **Receiving Antennæ or absorbing Circuits.**—We have hitherto chiefly paid attention to the processes by which oscillations set up in an antenna radiate electromagnetic waves, but we must at this stage consider more fully the action of an antenna or closed circuit as an absorber of electromagnetic radiation.

If we move a conducting wire through a field of magnetic force so as to cut across the lines we generate in it an electromotive force which is proportional to the magnetic force, to the length of the conductor, and to its velocity at right angles to the direction of the field. If the conductor is stationary, but if the magnetic force lines move across it, the same effect is produced. This action is the basis of operation of all dynamo electric machines. If, then, an electromagnetic wave falls on a conductor in such a manner that the magnetic component of the wave is transverse to the conductor, we may regard the movement of the wave as being equivalent to a cutting of this conductor transversely by lines of magnetic force. Hence, an alternating electromotive force is created in the antenna which is proportional to its length, and to the intensity of the magnetic force. The wave, however, possesses an electric force component, and this is at right angles to the magnetic component. If the magnetic component is at right angles to the absorbing antenna, then the electric component is in the same direction as the antenna, and as the wave passes over it both these components expend their energy in creating electromotive force in the antenna in the same direction. The wire in fact absorbs the electric component in the direction of its length, and is cut by the magnetic component transversely to its length. Both these operations contribute to create in the antenna an electromotive force. In a complete wave, the energy of the magnetic component is equal to the energy of the electric component, and hence both contribute equally to the production of the electromotive force.

It follows from this, and from the Law of Exchanges, that if a linear open circuit receiving antenna is placed at right angles to a similar radiating antenna, it will absorb nothing, and have no electromotive force created in it. If the antennæ remain in parallel planes, but if their directions are inclined at an angle θ, then the effect produced in the receiving antenna is equal to cos θ times that which would be produced if the antennæ were parallel. In optical language, a linear antenna emits a plane polarised wave, and the receiving antenna must be parallel to the electric component of that wave.

If the receiving antenna is a closed or partly closed circuit, then a third source of electromotive force exists. Consider the case of a closed receiving circuit, placed with its plane vertically to the earth and in the direction in which electromagnetic waves are passing over it, with their electric component perpendicular to the earth, and therefore their magnetic component perpendicular to the plane of the receiving circuit. As the waves advance, their magnetic component cuts through the vertical sides of the closed receiving circuit, and their electric component is more or less absorbed by the same sides. These actions contribute to the production of an electromotive force in the circuit, which is in one direction as the components pass over the near side, and in the reverse direction as they pass over the far side. But, in addition, if we consider the group of lines of magnetic flux in the wave which at any moment fill the closed receiving circuit and perforate through it, we shall see that the effect of the advancing waves is to cause a periodic change or alternation in the amount of magnetic flux thus perforating. If the closed circuit is rectangular and exactly half a wave length long, this action will produce the maximum effect, and we may adopt a phrase used in connection with the theory of induction motors, and call it the transformer effect of the wave, whilst the cutting of the sides of the receiving circuit by the moving lines of magnetic force is called the dynamo effect.

It will be seen on consideration that the relative magnitude of the electromotive forces set up by these three actions is capable of variation by many factors. Moreover, it gives the closed receiving circuit a directive power, or power of determining the direction in which the waves are travelling, not possessed by the simple open

vertical receiving antenna. Thus, for instance, if the closed receiving antenna has its plane perpendicular to the direction of the incident waves, it will not be affected at all.
at once to a means of determining this direction. We shall return to this matter in the last section of this chapter.

5. **The Practical Construction of Antennæ.**—The simplest form of radiotelegraphic antenna is a long vertical wire, A, which is suspended by an insulator at its upper end from a mast M, tower, or building (see Fig. 7). The lower end is in connection with one of a pair of spark balls S, the second of which is connected to a plate of metal E sunk in the earth. Owing to the fact that the electrical oscillations are confined to the surface of the wire, it is not desirable to employ a wire of large diameter. If a solid wire is used, it should not exceed 0·1 or at most 0·125 inch in diameter, that is, 2·5 to 3 mm., but it is generally better to employ a stranded wire, say, one made up of 7 No. 20 or 7 No. 22 wires ($\frac{7}{20}$ or $\frac{7}{22}$ S.W.G.).

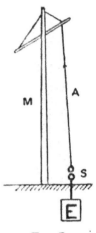

Fig. 7.

As regards material, tinned copper wire is most usually employed when stranded wires are used, but for solid wires aluminium may be used with advantage. The density of aluminium is 2·6 against 8·9 for copper. Hence, for equal bulks aluminium has only one-third the weight of copper. The price of aluminium is now about £80 per ton, and that of copper (in 1915) £70. Hence, for wires of equal length and diameter the cost will be proportional to the product of the density of the material and the price per ton. Accordingly, the aluminium wire will only cost about one-third of that of a copper one of the same bulk. The tensile strength, however, must be taken into account. That of commercial aluminium is from 26,000 to 40,000 pounds per square inch, and that of soft drawn copper is about 30,000 pounds per square inch. Alloys of aluminium are, however, now made with a density not exceeding 2·7, which have a tensile strength as great as that of copper.

In the case of high frequency currents, the electric conductivity does not much matter, as current is carried chiefly on the surface. The wire, however, must not be of iron, as the magnetic hysteresis of this iron would increase considerably the damping.

The experience of the author has shown that aluminium withstands the weathering action of the atmosphere very well. The chief precaution which must be taken in its use is to avoid the galvanic action which is set up when aluminium (which is highly electropositive) is brought in contact with other metals. The end of an aluminium wire should not be twisted up with a copper, brass, or iron wire if the junction is exposed to moist air, but insulated from it. Also aluminium wires are more difficult to solder effectually than copper wires, but these disadvantages are more than outweighed by the advantages of its use. We have only one-third of the weight to support with the same windage surface, and as aluminium wire can be obtained in coils of any reasonable length, there is no need to make joints in it. If the upper end is attached to a brass or copper ring, the latter should be wound over with indiarubber tape, to avoid a metal-to-metal contact between the aluminium and copper, which would cause the former to wear away by local galvanic action.

There is no necessity to put an insulating covering on the aerial wire, although indiarubber-covered stranded copper wire as used for electric light wiring is sometimes employed for antennæ.

A wire 100 feet in length and 0·1 inch in diameter has an electrical capacity of about 0·0002 microfarad or 180 electrostatic units, from which it will be seen that its energy storage in no case can be very large. Accordingly, when more capacity is required, we must either use several wires or else add metal surface at the top. In some of his early work Marconi employed as an antenna a long strip of galvanised iron wire netting, or else a single wire, with a cylinder of such wire netting placed at the top, or else a kite or balloon having its surface covered with tinfoil. The objection to the use of such *capacity areas* at the top of the wire is that they offer a greatly increased surface for the wind to act upon. Hence multiple antennæ are generally preferred.

6. **Multiple Wire Antennæ.**—Several wires may be arranged in many different forms to obtain an antenna of large capacity. The simplest method is the *double cone* antenna. In this case a light wooden star is formed by crossing a number of laths of wood like a star. Long wires of equal length are passed through holes in the ends of these laths, and tied together at their ends (see Fig. 8).

Two such wooden stars or crosses may be employed, and a cylindrical or fourfold antenna thus made (see Fig. 9). Otherwise, wires may be arranged in fan fashion (see Fig. 10), being joined at the upper end by a wire or rope, and bunched together at the lower ends; or they may be arranged conically.

In grouping together wires in this manner, whilst we increase the capacity relatively to a single wire, the increase in capacity is by no means proportional to the number of the wires. If two wires 100 feet long are hung up vertically parallel to each other, and about 4 or 5 feet apart, the total capacity is not double that of one wire, because the lines of electric strain proceeding from each wire to the earth are not then distributed symmetrically. Those of each wire disturb the distribution of the other. The nearer the wires are brought, the more we make their joint

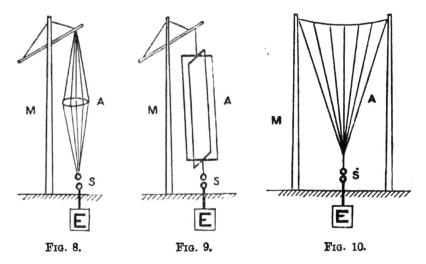

FIG. 8.   FIG. 9.   FIG. 10.

capacity less than the sum of their individual capacities when far apart.

Thus, if four wires 100 feet long and 0·1 inch in diameter are hung up vertically very far apart, their joint capacity would be about 0·0008 mfd., but if brought within 4 or 5 feet of each other, it would not be more than 0·0004 mfd.

As a rule, a few wires spaced fairly far apart are better than very many wires near together, as far as total capacity is concerned. The multiplication of the wires in an antenna has, however, another effect. It decreases the inductance of the antenna and also its high frequency resistance. It decreases, therefore, the time period of oscillation in so far as inductance is concerned, although on the whole there is generally an increase in the time period.

Its chief advantage is that it enables us to accumulate more energy in the antenna in virtue of the greater capacity. It is an advantage to increase this capacity without adding to the height of the wires, because an increase in height involves more cost in supporting them. One way of doing this is to carry the wire up vertically for a certain height, and then extend it horizontally. This may be done in one, two, or more directions, and we have a gallows-shaped, or T-shaped, or umbrella-shaped antenna.

M

This last form, with the radiating wires inclining downwards, is a favourite form, since it is easy to erect, and the wires themselves can act as stays for the central support. It is also a convenient form for the portable antennæ used for military radiotelegraphy (see Fig. 11).

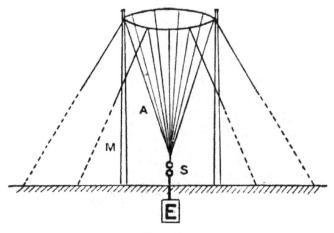

Fɪɢ. 11.

There are certain forms of antennæ which have the property of sending out electromagnetic waves more in one direction than others, and these are called *directive antennæ.* They will be considered in a later section.

For ship antenna, a special gaff or sprit is attached to a mast to give greater height, and a multiple antenna may be suspended from it, or, for some purposes, horizontal wires may be carried between masts, and vertical wires brought down from them from the middle or from both ends.

7. **Earthed and Non-earthed Antennæ.**—From what has been already stated it will be evident that in addition to classification into open and closed antennæ, we must distinguish between *earthed* and *non-earthed* antennæ. If, for instance, we stretch an insulated wire horizontally, cut it in the centre and introduce a spark gap, we construct a non-earthed horizontal antenna. When traversed by oscillations the lines of magnetic force are arranged round the antennæ with their planes vertical, and the electric force is in lines which lie in radial planes passing through the wire. If such a horizontal antenna is placed a little above the earth's surface, then at some distance from the antennæ in a direction perpendicular to the oscillator, the magnetic force is

vertical to the earth's surface and the electric force is horizontal. If we turn the above oscillator into a vertical position and place it at some distance above the earth's surface, the magnetic force will then be parallel to the earth's surface. In both these cases we construct what are called non-earthed oscillators. Supposing, however, in the last case the vertical Hertzian oscillator is partly buried in the earth so that one of the rods is completely underground, the spark balls being just above the surface, we then construct a *vertical earthed antenna*, and on considering the distribution of electric and magnetic force round it, it will be seen that the magnetic force is at all points parallel to the earth's surface, whilst near the earth's surface, at some little distance from the oscillator, the electric force is always vertical.

If one wire of the antenna is placed vertically and the other horizontally, the spark gap being at the angle where they meet, both wires being insulated, we have a form of non-earthed antenna, in which the horizontal wire is sometimes called the *balancing capacity*. This capacity may take the form of a sphere or metal cylinder also insulated from the earth (see Fig. 12). On the

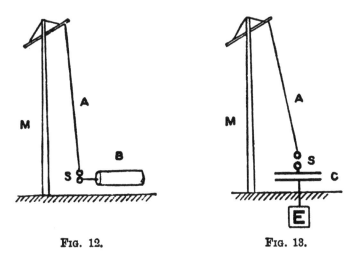

FIG. 12.   FIG. 13.

other hand, if a vertical earthed antenna is constructed, and if a condenser of large capacity is inserted between the spark balls and the earth (see Fig. 13), we do not in fact make a non-earthed antenna. A condenser can be traversed by a high frequency alternating current, and hence if the capacity is large enough to pass the same current as a dielectric current that would pass as a conduction current if the condenser were removed and the conductive earth connection restored, the presence of the condenser does not render the antenna in effect non-earthed.

It is well to bear in mind that generally speaking when we are concerned with high frequency currents a condenser acts as a conductor, whilst a large inductance acts as a non-conductor to these currents.

A form of radiator which is sometimes employed consists of a sheet of insulated metal placed close to the earth and connected by a vertical wire or wires with another plate elevated above the earth, the wire being interrupted by a spark gap placed near the lower plate (see Fig. 14). The real distinction to be made between these various forms of open circuit radiator is the mode of distribution of the lines of electric force before discharge. In the case of perfectly non-earthed or Hertzian radiators, the lines of electric force which start from one-half of it, extend through space and terminate in the other half, the two parts being separated by the air gap, and constituting the two plates, as it were, of a condenser.

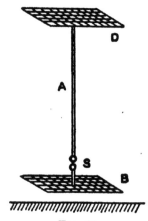

FIG. 14.

In the case of the conductively earthed or Marconi radiator, the lines of electric force before discharge stretch from the antenna and terminate on the earth's surface in its neighbourhood. In the case of antennæ comprising a balancing capacity of any form placed near the earth, there are a triple set of lines of electric force. One set extend from the vertical wire to the balancing capacity on which there is an opposite charge of electricity, others from the antenna to the earth, and a third set extend from the balancing capacity to the surface of the earth. If the balancing capacity is not removed a considerable distance from the earth, then during the oscillations we have rapid changes of potential of the earth in the neighbourhood of the antenna just as in the case of the conductively earthed or Marconi radiator.

The difference between the radiating effect of these forms of radiator has been much discussed. All practical experience shows that to produce telegraphic effects at a great distance, the lower ends of the radiating and receiving antennæ must be conductively connected to the earth, or what is equivalent to it must be connected to the earth through a condenser of large capacity; that is to say, currents of electricity must flow into and out of the earth in the neighbourhood of the antennæ. Also the radiating antenna must be so arranged that the lines of its magnetic force are parallel to the earth's surface, and the lines of its electric force at a distance from the antenna and vertical to it.

The reason for this is that the propagation of an electromagnetic wave over the surface of the conducting sea or land requires that the lines of· electric force should terminate on the earth's surface perpendicularly to it, or else that they should be detached from the oscillator in the form of complete loops with their planes perpendicular to the earth's surface. The magnetic force is then parallel to the earth's surface. If the oscillator is placed in any other position, say, horizontal, then the energy of its oscillations expends itself more or less in making induced or secondary currents in the earth's surface beneath it, and there is a corresponding diminution in the energy radiated. To produce the most effective radiation, the conduction current in the oscillator should be perpendicular to the earth's surface; it then creates rapid alterations of potential in the earth's surface beneath it, and detaches from itself lines of electric force and therefore radiates, but it is not then so situated as to enable it to expend its energy in creating induced or secondary conduction currents in the earth.

It is found, therefore, that for effective long distance radiation the antenna must either be vertical or have a considerable part of its length vertical, and that its lower end must be in such connection with the earth, that electricity can flow into and out of the earth out of and into the antenna very freely.

Over short distances it is quite possible to produce sufficient radiation for telegraphic purposes with a non-earthed antenna having a balancing capacity, but when long distances have to be covered the direct connection to earth must be adopted or else a condenser of large capacity interposed between the base of the antenna and the earth which is equivalent to an earth connection.

The practical construction of the "good earth" required will be considered in the chapter on Radiotelegraphic Stations.

As the law of exchanges necessitates an identity in the nature of the radiating and absorbing circuits, it follows that when employing earthed radiating circuits we must employ a similar circuit for receiving and absorbing the energy of the waves radiated. The undoubted advantage gained by the employment of earthed radiating and receiving antennæ, and the known fact that the distance at which radiotelegraphy can be conducted with a given energy expenditure depends upon the nature of the earth connection, and of the intermediate soil or surface whether land or sea, over which the waves pass, raises the important question of the true function of the earth in this matter.

The materials of which the surface crust of the earth is composed are very poor conductors of electricity for direct or unidirectional currents when perfectly dry. They owe their conductivity chiefly, if not altogether, to the water contained in them. Sea water, owing to the presence of salts in it, is a better conductor than rain or river water. Hence, the conductivity of the soil or sea is essentially electrolytic conductivity. For this reason damp soil has a high dielectric constant, since that of pure water is 80 compared with air as unity. The resistance per centimetre cube of water and soil, and their dielectric constants and specific resistances for direct or low frequency alternating currents are very roughly as follows :—

	Specific resistance in ohms per centimetre cube.	Dielectric constant k air = 1.
Sea water	100	80
Fresh water	100,000	80
Damp soil	10,000 to 100,000	5 to 15
Dry soil	1,000,000 and upwards	2 to 6
Dry rocks	practically insulators	—

Electric theory shows that the electric force outside and very near the surface of a good conductor must always be perpendicular to it, and also that the electric force must be zero inside the conductor. Hence, if the surface soil was as good a conductor as a metal, and we set up at any point on it a radiating earthed antenna, the electric component of the waves at or very near the earth's surface would be normal to the surface, and there would be no sensible penetration of the wave into it. There would be no absorption of the energy of the waves, but they would glide over the surface without other weakening than that due to the diffusion of the energy over a greater space. If, on the other hand, the soil was a perfect insulator there would be a penetration of the wave into the soil, but no loss of energy by absorption. The sea or soil is, however, a poor conductor or bad insulator, and this implies that there is a weakening of the electromagnetic wave moving over its surface by which some of the wave energy is frittered away as heat in the surface soil. This absorption reaches its maximum for a certain degree of specific resistance and dielectric constant.

This matter has been treated theoretically by J. Zenneck, who has given his results in the form of curves calculated for an assumed wave length of 1000 feet or a frequency of one million.

Supposing such long electric waves, which are of the wave length mostly used in radiotelegraphy, to travel over surfaces of various conductivities and dielectric constants, Zenneck has calculated the distance at which the wave amplitude would be reduced to $\frac{1}{2\cdot718}$ or to $\frac{1}{\epsilon}$ (where ϵ is the base of the Napierian logarithms) of the amplitude at the origin, apart altogether from reduction in amplitude due to the spreading of the wave energy over a larger area with increasing distance. His results are exhibited in the curves in Fig. 15, in which the ordinates repre-

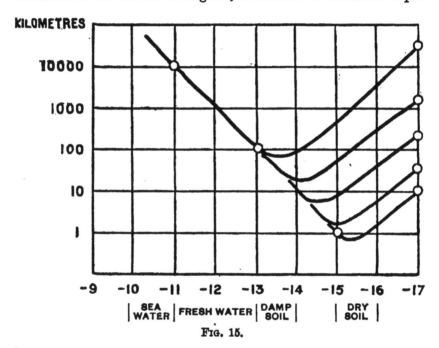

Fig. 15.

sent this critical distance in kilometres and abscissæ the logarithms of the specific resistance of the surface of the material over which the waves travel in ohms per centimetre cube. We have marked the range of resistivity corresponding to sea water, fresh water, and dry soil respectively, and it will be seen that there is an extraordinary reduction in the distance corresponding to a given reduction in the wave amplitude, with increasing resistivity of the surface over which the waves travel. It is this absorption which accounts for the well-known fact that radiotelegraphy is conducted with much more difficulty over dry land than over sea water.

In radiotelegraphy the radiating and receiving antennæ are

placed at the separating surface of two media, one the air, having an almost perfect non-conductivity and unit dielectric constant, and the other the earth or sea water, having a more or less imperfect conductivity and a dielectric constant greatly exceeding unity. The wave glides over the bounding surface, but partly penetrates into the soil or water, and on one side suffers absorption or loss of energy in so doing, and is thereby weakened. Although, however, the above facts give some reason for the greater difficulty of conducting radiotelegraphy over dry land than over sea, they do not account for the great improvement effected by making the earth connection at both ends. Radiating and receiving antennæ both connected to the earth respond at greater distances for the same energy expenditure than antennæ connected to perfectly insulated balancing capacities placed at some distance above the earth. In the former case the two antennæ and the earth virtually form one oscillator connected through a conductor of a certain capacity. If we may assume that the upper layers of the atmosphere are non-conducting then the capacity of the earth considered as a sphere in space is only about 800 mfs., or about equal to that of an Atlantic cable, and the addition to or subtraction from it of small quantities of electricity is therefore able to alter quite appreciably its potential relatively to some fixed zero. On the other hand, if the upper layers of the atmosphere have any appreciable conductivity the terrestrial capacity may be much greater.

Hence, if we have a conductor which is charged and discharged alternately from or to the earth, it is in fact giving or taking electricity to or from the earth, and therefore raising or lowering the earth's potential at that point relatively to some absolute zero. These sudden changes of potential at one point must be propagated over it, and an oscillation detector placed at a distant spot in the circuit of another syntonic antenna will detect these changes.

From this point of view the earthed radiating antenna and the earthed receiving antenna and the earth itself constitute one single oscillator, and rapid variations in the electric distribution at the radiator are felt and detected at the receiver in virtue of operations taking place in the crust of the earth and not entirely confined to the superincumbent dielectric.

It has been asserted by Sir Oliver Lodge that the radiation from a non-earthed antenna is less damped than is the case with earthed antenna, but the difficulty is to institute a comparison in which the initial energy storage is the same and all other circumstances identical except in regard to the connection or not to the earth.

The numerical values given on a previous page for the conductivity and dielectric constants of water and soil of various degrees of dampness are at best very rough, and all that they can serve for is to show us qualitatively the reasons for the greater facility of propagation of waves used in wireless telegraphy over sea than over land. It has been found by Dr. Austin and others that certain districts of the earth exercise an abnormal effect in increasing the energy loss of radiotelegraphic waves passing over them. This effect is shown in a weakening of signal strength. This weakening shows itself much more for certain wave lengths than for others. Thus in the above-mentioned case Dr. Austin found a certain district near Newport, U.S.A., over which waves of 1000 metres wave length suffered far more absorption than waves of 3750 metres length. Recent researches have shown that the true conductivity of such materials as compose the earth's crust is very different for direct or continuous currents and for alternating currents. The alternating current conductivity is much greater than the direct current conductivity. Again, it has been found that the conductivity of these materials for high frequency currents of such frequency as is used in radiotelegraphy is much greater than for ordinary low frequency alternating currents. Accordingly it appears that the earth crust materials, such, for instance, as slate, marble, granite, etc., are very much better conductors for these high frequency currents than would be inferred from any of the usual methods of measuring insulation resistance. The high frequency conductivity appears to reach a maximum for a certain frequency and then to decrease.

It has been found that radiotelegraphic signals can be received over long distances without any high or elevated receiving antenna provided the receiving apparatus has one terminal connected to a good earth and the other to some conductor having sensible capacity with respect to the earth. As far back as 1901 Marconi received signals at Poole sent from the Isle of Wight by connecting his receiver terminals to an earth plate and to a large zinc cylinder standing on a chair. Campbell Swinton has in the same manner employed a slightly insulated iron bedstead as a capacity, and the author has made use of a zinc dustbin to receive signals in London sent out from Paris.

These signals are much weaker than when a high elevated receiving antenna is attached, but the fact that they can be received at all seems to indicate that mechanism by means of which they are transmitted is not wholly an electric space wave in the æther, but partly of the nature of a surface wave of potential transmitted along the earth's surface materials.

8. The Establishment of Fundamental and Harmonic Oscillations in Open and Closed Circuits.—It is well known that a stretched string, such as a violin string, can not only vibrate as a whole but can divide itself into oscillating sections, having lengths respectively equal to one-half, one-third, and one-quarter, etc., of the whole length, in which case it emits notes of higher frequency than when vibrating in a single undivided length. The stationary points on the string are called the *nodes,* and the places of greatest amplitude of motion the ventral points, *antinodes* or *loops.* The skilful violinist can by touching the string lightly at certain places and bowing at other places thus cause a single string to vibrate in harmonics and emit a series of notes having frequencies 2, 3, 4, etc., times that of the fundamental note due to the vibration of the string as a whole. The same is true of the oscillations of the air in an organ pipe, as described in any book on acoustics.

In the case of electric circuits possessing capacity and inductance, we have a similar phenomenon. Thus we can set up electric oscillations in a linear Hertzian oscillator or Marconi antenna, called the *fundamental* oscillation. In this case there is no variation in potential at the centre of the Hertzian oscillator or at the earthed end of the Marconi antenna, and this point is therefore called a node of potential. At the free ends or upper end the variations of potential are a maximum, and these points are therefore called loops or antinodes of potential. The variations of potential increase from the centre or lower end to the free or upper end, and in the case of the Hertzian oscillator the charges at the two free ends are always opposite signs. Hence, we may represent these variations of potential by the distance of a dotted curve from a thicker line representing the oscillator or antenna (see Fig. 16 (*a*)). At the same time there are variations in the conduction current in different parts of

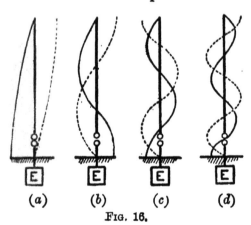

(*a*)   (*b*)   (*c*)   (*d*)

Fig. 16.

the oscillator or antenna which may be represented in the same way by a fine, firm line. At the centre of the Hertzian oscillator or the earthed end of the Marconi antenna the amplitude of the current is a maximum, and that point is called an antinode or loop of current.   On the other hand, at the free end, the conduction current is zero, and therefore these points are nodes of current. The current therefore increases gradually from the node to the antinode, and is not the same at all points on the antenna. Its amplitude at any point may thus be represented by the distance of a fine, firm line from a thick line representing the oscillator, as in the diagrams in Fig. 16 (*a*).

Again, we may set up oscillations in an open or closed circuit, which are called *harmonics* of the fundamental.   For example, in the case of the earthed Marconi antenna, we may set up a first harmonic oscillation in which in addition to the node of potential at the earthed end, there is another. node of potential at about one-third of the length of the rod from the open end, and a node of current at about one-third of the length of the rod from the earthed end, the variation of potential and current amplitude along the rod being represented by the fine dotted and firm lines, as in Fig. 16 (*b*).

A second harmonic may also be set up in which there are two nodes of potential in the rod in addition to one at the earthed end, and a corresponding distribution of current, the rule being that the earthed end must always be a node of potential, and the free or insulated end an antinode or loop of potential, whilst the earthed end is an antinode of conduction current and the free end a node of conduction current, nodes of current coinciding with the antinodes of potential, and *vice versâ* (see Fig. 16 (*c*), (*d*)).

In the case of a linear Hertzian or non-earthed symmetrical oscillator having the spark gap at the centre, there is at that point a node of potential and an antinode of current.   If we consider the nature of the distribution and the lines of electric force round a vertical earthed antenna, we see that in the case of the fundamental oscillation the lines of electric force stretch from the antenna to the earth in all directions round it, and when oscillations are excited by the discharge, these lines are detached as semi-loops of electric force with their feet on the earth.   When, however, harmonic oscillations are produced in the antenna, we have as it were a superposition of a complete non-earthed antenna, on the top of an earthed one, and we must therefore have a detachment not only of semi-loops but of complete loops of electric strain which move outwards into surrounding space, and together with the corresponding expanding circular lines of magnetic force which together constitute the radiation.

When a plain earthed antenna has electrical oscillations excited in it which are the first harmonic of its fundamental, the frequency of these oscillations is three times that of the fundamental oscillation, whilst the wave length is one-third. In the case of the second harmonic the frequency is five times and the wave length one-fifth of that of the fundamental.

In the case of a closed or nearly closed oscillatory circuit, we can also set up oscillations which are either a fundamental or a higher harmonic. Thus, in the case of a closed circuit consisting of a condenser and a circle of wire, with a spark gap opposite to the condenser (see Fig. 17), the fundamental oscillation involves a conduction current to and fro in the circuit with a node of potential at the centre of the spark gap and loops or antinodes of potential at the condenser plates. The condenser plates always

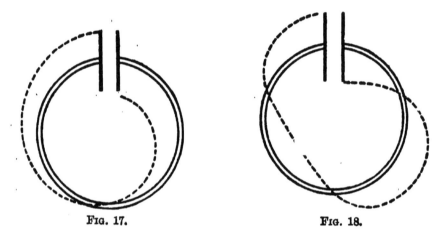

Fig. 17.   Fig. 18.

carry electric charges of opposite sign, so that the distribution of potential for the fundamental oscillation is represented in Fig. 17 by the radial distance of the dotted line from the thick black one representing the oscillatory circuit. The amplitude of the conduction current in the wire may be represented by the radial ordinate of another line, and this is a maximum at the spark gap, and has a minimum but not zero value at the condenser surfaces. In the first harmonic oscillation of such a circuit there are three nodes of potential and three antinodes, as shown by the radial ordinates of the dotted line in Fig. 18, and similarly, three places of minimum amplitude of the conduction current. It is, however, much more difficult to excite these harmonics when the capacity is concentrated or localised than when it is distributed throughout the circuit as is the case in an aerial wire or antenna.

This production of fundamental and harmonic oscillations in an antenna can be very beautifully exhibited by an experiment devised by G. Seibt, which has been improved by the author. A very long helix of fine silk-covered wire is made by closely winding in one layer the covered wire on a long round rod of ebonite. The helix constructed by the writer was nearly 2 metres or 80 inches in length, and 5 cms. or 2 inches in diameter. This helix is supported on insulators in a horizontal position about 2 feet above a table.

A closed oscillatory circuit is then formed of a condenser consisting either of one or more Leyden jars C_1 C_2 or of sheets of ebonite coated with metal and placed in oil and a variable inductance made as described in Chapter II. (see Fig. 19). Also a

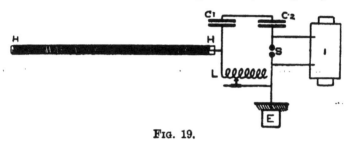

FIG. 19.

spark gap, S, must be provided in this oscillation circuit consisting of a pair of zinc or brass balls adjustable as to distance, and enclosed in a wooden box. These balls are connected to the secondary terminals of an induction coil, and one of them is connected to the earth E. The helix is connected to this oscillatory circuit as shown in Fig. 19. The first step is to so adjust the capacity and inductance in the oscillatory circuit as to tune it to the natural time period of the helix. In the case of the helix made by the author, the helix consisted of 5465 turns of wire on a rod 215 cms. in length, and its natural time period of oscillation was nearly $\frac{1}{200000}$ of a second or about 5 microseconds. The condenser used in the oscillation circuit had three sections of capacity 1461, 2887, and 5835 micro-microfarads respectively. The inductance used could be varied from 5000 to 120,000 absolute units or from 5 to 120 microhenrys. The oscillation circuit was in tune with the helix when the former was composed of an inductance of 110 microhenrys and a capacity of 5885 micro-microfarads, since these last constants correspond to a frequency n, where,

$$n = \frac{5 \cdot 033 \times 10^6}{\sqrt{0 \cdot 005885 \times 110000}} = 0 \cdot 197 \times 10^6$$

which is equal to a time period of nearly 5 microseconds.

When oscillations are set up in the condenser circuit by connecting the spark balls to an induction coil they will excite other oscillations in the helix, and these create around it an electric field which may be detected by holding near the helix a vacuum tube preferably filled with Neon. If this tube is held in various positions, it will be found to glow with increasing brilliancy as moved from the condenser end towards the free end of the helix, owing to the gradual increase in the potential amplitude as the free end is approached. If, however, the inductance and capacity in the condenser circuit are altered so as to make the frequency 3, 5, or 7 times that of the fundamental oscillation of the helix, it is found that a state of vibration is set up on the helix in which there are nodes of potential near which the vacuum tube does not glow. Thus we can set up a state in which there is one such node of potential about one-third of the length of the helix from the free end, and likewise states in which there are 2, 3, etc., such nodes.

9. Modes of Exciting Oscillations in an Open or Closed Radiating Circuit.—We have next to consider the various modes of exciting oscillations in a radiating circuit. The simplest method is the production of damped oscillations by inserting a spark gap in the circuit. Thus, to excite such oscillations in an open earthed plain Marconi ærial, a spark gap is inserted near the earth. The spark balls are connected to an induction coil or transformer, and the antenna is therefore charged either intermittently or alternately, and discharged across the spark gap. In this case the charge is limited by the capacity of the antenna and the voltage which the coil or transformer can give, and this process results, as we have seen, in the production of highly damped oscillations. The energy stored is necessarily small, and hence is soon frittered away (see Fig. 20).

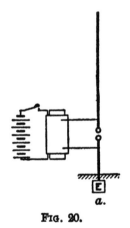

Fig. 20.

In the case of a closed circuit, including a condenser of large capacity, a much larger energy storage is possible, and since the circuit is a worse radiator than an open circuit, the trains of oscillations, and therefore of the waves emitted, are less damped, and contain more oscillations per train. This method of excitation by a spark gap is called the *direct excitation*.

The second method is a method of *direct coupling*. In this case oscillations are excited in a closed circuit containing a condenser inductance and a spark gap, and since the circuit is a bad radiator and the condenser can have a large capacity, a considerable storage of energy is possible. To this circuit at one point is connected an antenna or good radiating circuit (see Fig. 21), and another point on the closed circuit is connected to the earth. The antenna or open circuit has its own natural period of vibration like that of the closed circuit, and the two must therefore be syntonised together. This is most easily achieved by inserting an inductance coil between the antenna or open circuit and the earth, and making a variable part of this inductance coil by means of a movable contact the inductance in the closed circuit (see Fig. 21). In this manner the oscillations of the two circuits can be varied until they are equal, and this equality can most easily be discovered by connecting a hot wire voltmeter across one or two turns of the

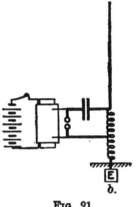

FIG. 21.

inductance coil near the foot of the open circuit and then varying the inductance or capacity in the open and closed circuits until this hot wire voltmeter gives its maximum reading. The two circuits are then said to be tuned together.

With such an arrangement we provide a much larger storage of energy than is the case in the direct method of excitation, since we associate with a feebly radiative closed circuit of large energy storage power a good radiative or open circuit. The method of direct excitation may be compared with a thermal radiator, in which a small quantity of a hot fluid, say water, is enclosed in a metal vessel covered with lamp black. The water then cools quickly for two reasons : first, because the surface of the vessel is a good thermal radiator; and secondly, because there is only a small store of liquid to cool. If, however, we were to connect this vessel with a reservoir made of polished metal containing a larger store of the hot fluid, then by the continuous circulation through the small but good radiative vessel of the large mass of hot fluid from the other the radiation could be maintained for a much longer period.

A third method of exciting the oscillations in an open radiative circuit is by the method of *inductive coupling*, using an oscillation transformer. In this case, we insert in the open radiative circuit one coil of a transformer comprising two interwound coils of wire, the second coil forming the inductance of the closed oscillatory circuit. Hence, when oscillations are excited in the latter by means of a spark gap, this will induce other oscillations in the open or radiative circuit, provided that the two circuits are brought into ·tune or syntony with each other (see Fig. 22). For this purpose both circuits must be provided with variable inductances, as was first done by Marconi, so that by variation of the inductance in the open circuit and variation of the capacity or the inductance in the closed circuit, the two circuits may be brought into tune with each other. We may discover when this is the case by connecting a hot wire voltmeter over one or two turns of the inductance placed in the open radiative circuit and then altering the inductance in one or both circuits, or the capacity in the closed circuit until the reading of this voltmeter becomes a maximum. The two circuits are then tuned.

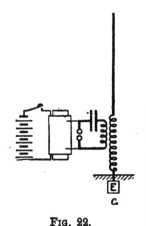

Fig. 22.

Unless this is done, the oscillations in the closed circuit will produce very little effective radiation in the open circuit. The two circuits, however, may be so adjusted that the time period of one circuit is a harmonic of that of the other, in which case they will operate as if coupled for fundamental oscillations.

There is a fourth method of connection called the *electrostatic coupling*, which however is not frequently employed. In this case, the open or radiative circuit terminates in a plate which is placed in contiguity to one of the plates of the condenser in the closed circuit. An arrangement employed by Hertz and also by Lecher makes use of this method of coupling. Oscillations are excited in an open or Hertzian oscillator, consisting of two plates at the ends of two rods, these rods being in one line and separated by a spark gap. In apposition to the two plates of the Hertzian oscillator are placed two other plates which are respectively connected to two long wires. These long wires must have their capacity adjusted by varying their length until they are in syntony with the primary circuit in which the oscillations are set up, and these last will then create secondary oscillations in the long wires, which will attain their maximum amplitude on exact syntonisation.

The most usual ·modes of connecting the good radiating circuit to the energy-storing circuit are by the direct or inductive coupling. The latter method has the great advantage that the closeness of the coupling can be ·varied, and hence also the nature of the oscillations set up in the secondary circuit. We have already shown that when two circuits having the same time period of oscillation are inductively connected, then, on establishing free oscillations in one circuit, oscillations of two frequencies are set up in both circuits, and that if the coupling is close these two frequencies are well separated, but if the coupling is loose they are merged into one. Hence, if we desire to set up in an antenna induced oscillations of one definite period the antenna must be connected in series with the secondary coil of an oscillation transformer, the primary coil of which is in the closed or condenser circuit, and the two coils must be well separated. If, however, the coupling is close we then have oscillations of two frequencies set up in the antenna and waves of two wave lengths radiated.

It can be proved by experiment and by theory that in the last case the wave train of largest wave length is the least damped, and the train of shortest wave length is most damped or has fewest oscillations. If the closed and open circuits respectively when separated have decrements δ_1 and δ_2, then when coupled together with a coupling coefficient k Wien has shown that the decrements of the two waves radiated by the open antenna will be D_1 and D_2, such that

$$D_1 = \frac{\delta_1 + \delta_2}{2\sqrt{1 - k}}$$

$$D_2 = \frac{\delta_1 + \delta_2}{2\sqrt{1 + k}}$$

Thus, for instance, in an experiment made with a coupled antenna at University College, London, the open antenna had an oscillation constant of its own of 7·0, and hence a frequency $n_0 = \dfrac{5 \times 10^6}{7} = 0.72 \times 10^6$, corresponding to a wave length $\lambda_0 = 1400$ feet.

This is the wave length of the wave which would have been emitted if this antenna had been used as a plain self-excited antenna. Its decrement δ_2 when so used was found to be 0·175. It was then coupled to a closed circuit containing a condenser of capacity 0·025 microfarad and an inductance of 2 microhenrys, and had therefore a natural frequency of 0·71 $\times$ 10⁶. It had a decrement $\delta_1 = 0.09$ when the spark gap in the circuit had a length of 2 mm. Accordingly, when the two circuits were coupled with a coupling coefficient $k = 0.5$ waves of two wave lengths were emitted from the antenna, viz.

N

$$\lambda_1 = \lambda_0\sqrt{1 + k} = 1400 \times 0.7 = 980 \text{ feet}$$
$$\lambda_2 = \lambda_0\sqrt{1 - k} = 1400 \times 1.224 = 1714 \text{ feet}$$

and these had decrements D_1 and D_2, such that

$$D_1 = \frac{0.09 + 0.175}{2 \times 0.7} = 0.18$$
$$D_2 = \frac{0.09 + 0.175}{2 \times 1.224} = 0.11$$

The longer wave of 1714 feet has the least damping, and therefore the longest train of waves.

In the case of the direct coupled antenna there is also an emission of waves of two wave lengths when the open and closed circuits are syntonised. In this case, however, the difference between their wave lengths depends upon a coefficient ρ, which is the square root of the ratio of the capacity c of the antenna with respect to the earth to the capacity C of the condenser in the closed circuit, and the wave lengths of the two waves emitted are given by

$$\lambda_1 = \lambda_0\sqrt{1 + \rho}$$
$$\lambda_2 = \lambda_0\sqrt{1 - \rho}$$

where $\rho = \sqrt{\dfrac{c}{C}}$

Hence, if c is small compared with C, waves of only one wave length are radiated.

This is usually the case, and the mode of direct coupling is therefore one which is very generally employed.

10. **Appliances for giving Direction to Electromagnetic Radiation. Directive Antennæ.**—We have seen that in the case of a vertical open circuit antenna the radiation is necessarily symmetrical in all directions, and is, therefore, equally detected by receiving antennæ placed at the same distance round it in all azimuths.

A problem which presented itself very soon in connection with radiotelegraphy was the limitation of this uniform all round radiation to a certain direction. It was obvious that some means was required to effect that which a lens or mirror effects in the case of light. Hertz had shown that electromagnetic radiation of short wave length could be reflected by metallic mirrors, and followed the laws of reflection of light. By means of parabolic mirrors he thus concentrated electromagnetic radiation into a beam. This, however, can only be done if the wave length of the radiation is not large compared with the dimensions of the mirror. Thus, for instance, Hertz placed two metal rods, each about a foot in length, in line with each other, and placed them both in the focal line of a parabolic cylindrical mirror. At a distance he placed another similar mirror with a receiving antenna in its axis, and, on setting up oscillations in the transmitting oscillator, he was able to direct a beam of electromagnetic radiation on to the other mirror and concentrate it in the focal line and hence detect it. Marconi was successful in projecting electromagnetic radiation by this means for a distance of about 2 miles when using waves of short wave length. But this can only be done by the aid of mirrors when the dimensions are comparable with the length of the waves employed. In radiotelegraphy, however, the wave length of the waves now employed is from 500 or 1000 up to 10,000 feet or more in length, and hence there is no possibility of constructing mirrors of sufficient dimensions to concentrate such radiation in any required direction.

A new line of investigation was, however, opened up by the observation that the radiation of a sloping antenna, and particularly that of a vertical loop or closed circuit radiator, was not symmetrical. In the case of the closed circuit it is greater in the plane of the loop than at right angles to it. Some preliminary investigations concerning this phenomenon were made by Sigsfield, Braun, Zenneck, Strecker, Slaby, Garcia, de Forest, and others, on the non-symmetry of radiation of inclined open antennæ, and Stone and de Forest, noting also cases of asymmetry in receiving antennæ, suggested means for locating the direction of an electromagnetic wave. But although claims were made for arrangements said to be effective, these various researches were not pressed to such logical issue as to disclose any definite scientific principle, whilst in some cases results said to have been obtained are clearly in contradiction to well ascertained facts. The problem of locating the direction in which an incident wave was arriving seems to have first attracted attention. If two vertical antennæ are erected on a plane at a distance apart equal to half the wave length of electromagnetic waves travelling over that plane, which have their magnetic force horizontal and electric force vertical, then their action upon these two antennæ will depend upon the direction or propagation of the waves. If, for instance, the waves are travelling in a direction parallel to the plane in which the two antennæ are placed, oscillations will be created in these two receiving antennæ which are opposed to one another in phase. If, however, the wave is travelling in a direction perpendicular to the plane containing two vertical antennæ, then the oscillations set up in them will be coincident in phase. If, then, these two vertical antennæ are insulated from the earth, and horizontal wires are brought near the earth from their base to a middle point and then earthed at the middle point, it is possible to make the oscillations propagated along these horizontal wires combine together at their junction in their action upon some form of oscillation detector made as described in the next chapter, so that when the oscillations are of the same phase, the oscillation detector is affected, but when the oscillations in the antennæ are opposed in phase, the oscillation detector is not affected. If, then, we could move round the two antennæ into various positions, keeping them half a wave-length apart, we could ascertain the direction in which the waves are travelling by ascertaining how the antenna must be placed to produce the greatest effect.

It will be obvious, however, that when dealing with waves of a thousand feet or more in wave length, this movement, although possible in idea, is not practicable in fact, and accordingly the method, although theoretically correct within certain limits, fails to give a solution of the problem for practical purposes.

The first real solution of the problem came from an observation made by Marconi that if an antenna is bent so as to have a short part of its length vertical and the greater part of its length horizontal, the lower end of the vertical part being connected to the earth and the outer end of the horizontal part being insulated, then such an antenna provides the means for producing a non-symmetrical radiation, and also for detecting the direction in which electromagnetic waves are passing over it (see Fig. 23). He found

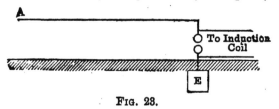

FIG. 23.

that such a bent antenna emits a less intense radiation at any given distance in the direction in which the free end points than in the opposite direction. Also, since the law of exchanges holds good for electric radiators, this form of bent antenna receives or absorbs best electric waves which reach it from a direction opposite to that in which the free end points. Hence, two similar bent antennæ when set up back to back, that is, with their free ends pointing away from each other, form a system of radiator and receiver which has a greater range in that position than in any other at the same distance, and hence has directive qualities not possessed by the ordinary vertical antennæ. Although the full explanation of this phenomenon requires the application of somewhat advanced mathematical analysis, it is not difficult to give a general explanation of it in non-symbolic language. Consider, for example, a square or rectangular circuit A, B, C, D (see Fig. 24),

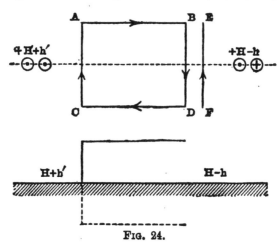

Fig. 24.

ın which electric oscillations are taking place. The magnetic field of such an oscillator consists of closed lines which embrace the circuit of the oscillator. These lines are all perpendicular to the plane A, B, C, D in that plane, and if the current is in any instant going round in the same direction to the hands of a watch, that is, in the direction A, B, D, C, the lines of magnetic force in the included space are proceeding away from the reader, and returning on all sides outside the area towards the reader.

If we represent the section of these lines of force by little ɛircles, and indicate that the line is coming towards the reader by a dot put in the centre, and that it is moving away from the reader by a cross, then we can represent the section of a pair of such lines of magnetic force outside the oscillator returning back on both sides in the equatorial line by the two little circles marked + H, in which the magnetic flux is towards the reader. On the other hand, if we consider a simple open antenna EF of the same height as the side of the rectangle BD, and consider the nature of the magnetic force round it when a current is flowing upwards in it, it will be seen that these lines are circles lying in planes at right angles to the antenna, and that the sections of these lines in that plane may be represented by the little circles $+ h'$ and $- h$ marked respectively with a dot and a cross. If, then, we suppose the open and closed circuits to be placed so that the open one is in close contiguity to one side of the closed one (see Fig. 24), and that the oscillations in these parts of the two circuits in contiguity are always in opposite directions, then it is quite easy to see that the field due to the open circuit antenna will assist the field due to the closed circuit antenna on the left-hand side, but tend to weaken it on the right-hand side. So that if we call the field due to the open antenna on the one side h, and on the other side h', the result in the field due to the combined open and closed antennæ will be $H + h'$ on the left-hand side, and $H - h$ on the right-hand side.

We can now imagine the two oscillations in the continuous wires BD, EF which are opposed in direction to annihilate each other, and the result is that we are left with a bent antenna as in the lower Figure 24, in which if oscillations are set up we are able to produce a field which is non-symmetrical, being greater on the side away from which the open ends point. Such an antenna is called a bent antenna, and if we imagine it half buried in the earth, the surface of the earth being a plane of zero potential, it produces the same effect above the earth's surface as one-half of a complete double bent antenna. It follows, then, that an earthed antenna partly vertical and partly horizontal must produce a non-symmetrical radiation.[1]

Marconi made this discovery experimentally as follows :— Setting up at some place a bent antenna as above described, he took observations of the strength of the field, that is, of the intensity of the radiation by means of a vertical receiving antenna placed at equal distance but in various directions around the bent antenna. Marking off then on a polar diagram of radial lines (see Fig. 25) the intensity of the radiation in different azimuthal directions, he obtained a closed curve something like a figure 8 with two unequal loops, the radii of this curve representing the intensity of the radiation for various angular directions round the bent transmitter. It will be seen that the radiation is greatest in one direction, and that is the direction away from which the free

---

[1] The above almost self-evident explanation of the action of a bent antenna has, however, not been accepted by German writers, although it is essentially confirmed by the experiments of Bellini and Tosi described below. J. Zenneck has advanced another theory in which he states that as far as regards the bent receiver antenna, the asymmetry depends on the alternating field of the transmitter being inclined to the vertical, and having therefore a horizontal component. See *Science Abstracts*, Vol. II. B., abs. 705, June, 1908.

end of the bent radiator points.   It is also a minimum in another direction approximately 110° from the maximum direction, and it has a secondary or intermediate minimum 180° in the opposite direction, that is, in the direction in which the free end of the bent antenna points.   The shape of this curve can be fully accounted for theoretically by assuming as above that the bent antenna is a combination of a closed or magnetic oscillator and an open or electric oscillator.

A large number of observations were obtained by Marconi with bent transmitting antennæ and vertical or open receiving antennæ, and also with vertical or symmetrical radiating antennæ and bent receiving antennæ placed in various relative positions, and these observations all confirm the statement

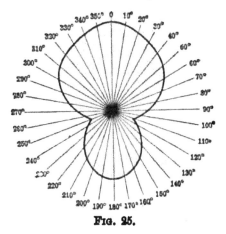

Fig. 25.

made above that an antenna which radiates best in any one direction absorbs best as a receiving antenna, waves which are coming from that direction, and also that when an antenna is constructed which is partly vertical and partly horizontal, the radiation is non-symmetrical, being greater in some directions than in others. Marconi's observations were made with radiating and receiving antennæ from 30 to 45 metres in length, separated by distances varying from about 250 metres to 600 or 700 metres, and he then found that for the same distance between the antennæ the intensity of the radiation, as measured by a thermal or magnetic oscillation detector, was sometimes as much as four times greater in the direction away from the free end of the bent radiator pointed than in the same direction.   The wave length of the waves used in his experiments was about 150 metres, and hence the maximum distance at which experiments were carried out was only about four or five wave lengths. Practical experience, however, shows that the same directive qualities exist at very much greater distances, but theory points to the fact that at extremely large distances the asymmetry tends to vanish, and that any bent oscillator, however arranged, has no asymmetry of radiation for very large distances. In one experiment he employed a horizontal wire 100 metres in length, placed at a slight distance above the earth's surface, and connected at one end through a spark gap with the earth. Such a transmitter sent out waves approximately 500 metres in length. The receiving antenna was a vertical wire 8 metres in length, tuned to the period of the transmitter by means of a syntonising coil and connected to the earth through a magnetic oscillation detector (see next chapter). The signals were quite distinct at 16 kilometres when the horizontal part of the radiator pointed away from the receiver, but only very weak at 10 kilometres when the free end of the transmitter pointed towards the receiving wire, and quite undetectable at 6 kilometres when the free end of the transmitter pointed at right angles to the line joining the transmitter and receiver.

Again, at Clifden, Connemara, Ireland, by means of a horizontal conductor 230 metres in length as a receiving antenna, and connected into the earth through a magnetic oscillation detector, Marconi found it possible to receive with clearness all the signals transmitted from the Poldhu station at a distance of 500 kilometres, provided that the free end of the horizontal receiving antenna pointed directly away from the direction of Poldhu, whilst no signals at all could be received if the horizontal wire at Clifden made an angle of more than 35° with the line of direction of Poldhu. Furthermore, he found that he could receive signals from the Admiralty Station on the Scilly Isles at Mullion in Cornwall, a distance of 85 kilometres, by means of a horizontal receiving antenna 50 metres in length placed 2 metres above the ground, one end of the wire being connected to the earth through a magnetic oscillation detector provided that the free end of the wire at Mullion pointed away from the Scilly Isles, but that no signals could be received if the horizontal portion was swivelled round so as to make an angle of more than twenty degrees with the line joining Mullion with Scilly.

Also by means of a horizontal wire 60 metres in length, supported 2 metres above the ground and being connected at one end to the earth through a magnetic oscillation detector, Marconi was able to locate the direction of an invisible ship sixteen miles away, sending out electromagnetic waves, by noticing the direction in which the free end of the horizontal receiving antenna had to be placed in order to make the signals most strong. This direction was a direction opposite to that from which the waves were arriving (see Fig. 26).

Some experiments of the same kind were made by the author in the same year. A vertical radiating antenna was employed consisting of a single wire which could be bent over at various heights from the ground, so as to make a bent antenna partly vertical and partly horizontal, the ratio of the horizontal to the vertical lengths being varied at pleasure. A vertical receiving antenna was employed at distances varying between 80 to 150 feet, and in the receiving antenna a hot wire oscillation detector of the thermoelectric type (see Chapter VI.), devised by the author, was employed to measure the R.M.S. value of the current created in the receiving antenna. The transmitting antenna had its horizontal part swivelled round in various directions at intervals of 15°, and in the several positions the current created in the receiving antenna was measured, the oscillations being excited in the transmitting antenna by means of a spark gap of constant spark length. The total length of the transmitting antenna was 20 feet, and the height of the receiving antenna was the same length.

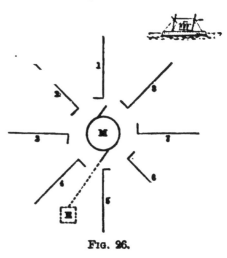

FIG. 26.

The following table shows the current in the receiving antenna in arbitrary units for each position of the horizontal part of the transmitting antenna

Radiation from a Bent Earthed Transmitting Antenna 20 feet in total length. Receiving Antenna vertical and 20 feet high. Distance between receiver and transmitter 138 feet.

Length in feet of vertical part of Transmitter	5	4	3	2	1
Length in feet of horizontal part of Transmitter	15	16	17	18	19
Radiated wave length in feet	100	100	105	106	110
Azimuth of horizontal part of Transmitter in angular degrees.	Current in the receiving Antenna in arbitrary units.				
0	100	100	100	100	100
15	98	97	94	92	93
30	92	85	96	83	75
45	82	79	79	77	67
60	78	74	70	71	58
75	77	67	59	56	45
90	72	66	57	52	48
105	71	65	57	46	41
120	70	66	62	53	49
135	72	64	60	54	48
150	73	80	58	67	59
165	70	74	56	69	60
180	82	69	64	63	68

These observations clearly confirm Marconi's observations that the radiation from a bent antenna is unsymmetrical, being greatest in a direction opposite to that towards which the free end of the antenna points. It was also found that by bending down the free end towards the earth, as in Fig. 27, the radiation became still more unsymmetrical, as shown by the polar curve in Fig. 27,

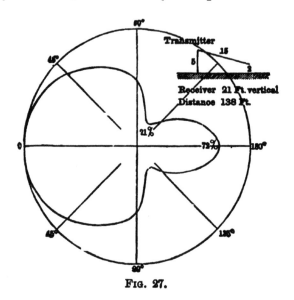

Fig. 27.

in which the radii represent the strength of the currents in the receiving antenna corresponding to various relative positions of the horizontal or inclined part of the transmitting antenna. It will be seen from Fig. 27 that the tipping down of the horizontal part causes nearly the whole of the radiation to be sent out towards that side opposite to which the free end points.

Another entirely different method of giving direction to electric waves has been devised by F. Braun, which depends upon the interference of electric waves travelling in the same direction but different in phase. In Braun's method, three simple vertical wire antennæ are set up in positions corresponding to the angular points of an equilateral triangle, and oscillations are created in these antennæ which differ from one another in phase. These oscillations with phase differences were produced by a method devised by Papalexi and Mandelstam. By these arrangements it is possible to cause the waves emitted by the free antennæ to combine together and assist one another in certain directions, but to neutralise one another in certain other directions. It is well known, for instance, as described in books on Optics and Acoustics, that waves of light or waves of sound can in this way interfere, so that two light waves may actually destroy one another and produce darkness, and two sound waves neutralise each other and produce silence. This effect is called the interference of waves. Braun found that by a proper arrangement of the antennæ and adjustment of the phase difference, the radiation of the three antennæ could be combined together in a certain region out of the whole azimuth of 360°.

The experiments were carried out on a large open space near Strasburg. Wooden poles 20 metres high were planted at the corners of an equilateral triangle whose sides were 30 metres long. Antennæ wires each approximately 33 metres long terminated in wire netting stretched parallel to the ground and at a small distance above it. These constituted the balancing capacities. In the centre of the triangle an observation hut was constructed from which the wires ran out horizontally to the masts at a height of $2\frac{1}{2}$ metres above the ground. At a distance of 1300 metres a receiving station was constructed and a receiving wire erected attached to a pole 20 metres high. In the circuit of this receiving wire was placed a hot wire oscillation detector (see Chapter VI.), by means of which the current in the receiving wire could be measured.

In a number of the experiments the oscillations in two of the transmitting antennæ were of the same phase, but differed from these in the third antenna by a definite amount, say, by 100°, this definite difference of phase being secured by the method of producing multiple spark discharges due to Mandelstam and Papalexi. The amplitude of the oscillations in the two antennæ in the same phase was half that in the third antenna. Under these conditions, if observations are taken of the current in the receiving antenna at equal distances, but in different azimuths round the triple transmitter, it is found that in one direction the radiation is a maximum, and in the opposite direction it is nearly zero, varying in accordance with the radii of a polar curve, as shown in Fig. 28.

The method, although ingenious, has not the simplicity and practicality of the bent receiving and transmitting antennæ employed by Marconi.

Another very ingenious system of directive radiotelegraphy has been devised by E Bellini and A. Tosi. They employ a nearly closed circuit transmitting antenna, consisting of two aërial wires suspended from one mast, the upper ends being insulated and the lower ends brought into a signalling house, the wires being

stretched out, as shown in Fig. 29, so as to give them the form of
a triangle.   If oscillations are set up either by the direct coupled

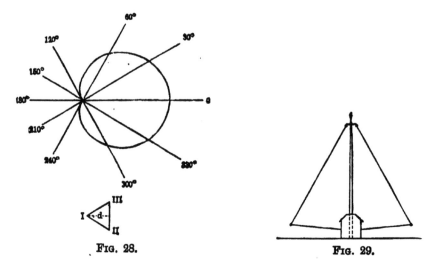

FIG. 28.   FIG. 29.

or inductive method, radiation takes place from this nearly closed
antenna which is not symmetrical, and is greatest and equal in the
two directions in the plane of the antenna and zero at right angles

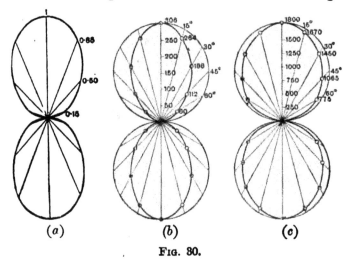

(a)   (b)   (c)

FIG. 80.

to that plane; in other directions varying in accordance with the
radii of a figure of 8 polar curve, as shown in Fig. 30 (a), (b), (c).
Fig. 30 (a) shows a theoretical curve of which the radii in various

directions are proportional to the energy sent out in these directions by a closed circuit oscillator with plane perpendicular to the paper, and in the direction of the maximum radius. Fig. 30 (b) shows a curve obtained by actual observations, employing the Duddell thermoammeter (see chapter VI.) to measure the energy radiated in various directions. Fig. 30 (c) is a similar polar curve, the radii of which denote magnetic field strength in various azimuths at equal distances from the transmitter.

These inventors employed a transmitting antenna and a receiving antenna of the same form. When used as a receiving antenna the oscillation detector, of whatever type it may be,

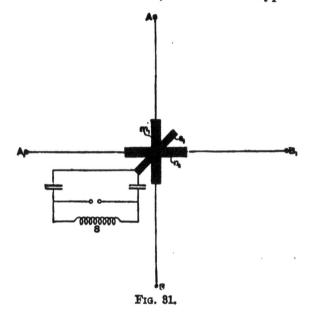

Fig. 31.

is placed at the centre of the lower or horizontal side of the triangle. When such a circuit is used for reception the intensity of the oscillations created in it by the incident waves is a maximum when the plane of the circuit coincides with the direction of the waves, and zero when it is at right angles to it. Such a circuit may be employed, therefore, to discover the direction in which its waves are travelling which fall upon it by swivelling round the circuit into various positions around its vertical axis, but Bellini and Tosi prefer to construct and erect two such circuits at right angles to one another at each station. Each of these circuits contains in its lower part a coil which can be acted upon inductively or can act inductively upon another circuit placed in an intermediate position, which last circuit either contains the oscillation producing arrangement, if it is a transmitter, or the oscillation detecting arrangement, if it is a receiver. The arrangement is shown in Figs. 31 and 32, the pair of closed

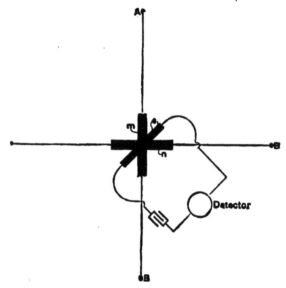

Figs. 29, 30, 31, 32 are reproduced from " Electrical Engineering " of November 14, 1907, by permission of the proprietors.

FIG. 82.

antennæ forming the transmitting and receiving arrangement respectively are placed with their planes at right angles, and the coils to be acted upon inductively, which are inserted in their lower portions respectively, being shown in plan and therefore at right angles. A third coil, which forms as it were the primary or secondary circuit of an oscillation transformer, is placed close to and within the other two coils just mentioned, and in the transmitter this last-named coil is connected respectively with the condenser and a spark gap, and in the receiver with an oscillation detector.   The coil in which the oscillations are either set up or are detected is capable of being swivelled round so as to be parallel with either of the coils contained in the circuits of the pair or closed antennæ. Supposing now the waves are incident on the receiving arrangement coming from a certain direction.   In order to determine that direction, all that it is necessary to do is to swivel round the secondary coil in direct connection with the oscillation detector so as to place it parallel to one or other of the primary coils inserted in the closed circuit antennæ or in some intermediate position. Some position will then be found in which the indications of the oscillation detector are a maximum, and when that is the case, the waves must be falling on the compound antennæ in the direction of the plane of the secondary coil attached to the oscillation detector. In the same way, to send out radiation, which is a maximum in any given direction, the coil in which the oscillations are being produced is swivelled round so as

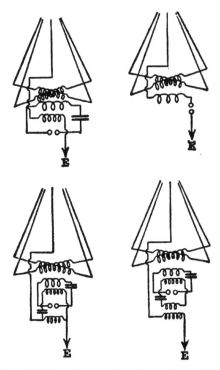

Reproduced from " Electrical Engineering " by permission of the proprietors.
FIG. 88.

to be parallel to one or other of the secondary coils inserted in the circuits of the two closed circuit antennæ, and the radiation will then be a maximum in the direction in which that primary coil points.

Experiments with this system showed that good results could be obtained with an expenditure of less than 500 watts between Dieppe and Havre (55 miles overland) and Dieppe and Barfleur (110 miles over sea). The angles between the stations, Dieppe-Havre-Barfleur is 23°, but the Dieppe-Barfleur transmission did not affect the Havre, nor did the Dieppe-Havre transmission affect Barfleur. The height of the antennæ was 48 metres, the wires being 60 metres long at the base and 60 long in the inclined side, forming an equilateral triangle, each side of which was 60 metres in length. It was also found that this closed circuit system was more proof against disturbances from atmospheric electricity (to which further allusions will be made in Chapter VII.) than the system employing open circuit antennæ.

The great advantage of the methods of Bellini and Tosi is

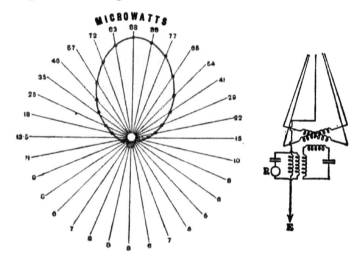

Reproduced from " Electrical Engineering " by permission of the proprietors

Fig. 34.

that no movement of the antennæ themselves is required to locate the direction of the radiant centre or to give direction to the radiation. The only movable part is a small coil which acts upon, or is acted upon, by the fixed antennæ. This arrangement for locating the direction of a station is called a *radiogoniometer* by its inventors, and promises to be of considerable use in connection with radiotelegraphy. By employing a single vertical antenna in combination with two nearly closed antennæ at right angles to each other (see Fig. 33), Bellini and Tosi have been able to confine the radiation of this compound antenna entirely to one side ; so that the polar curve of radiation is represented by a cardiod (see Fig. 34), the antenna being situated at the cusp.

It may be noted also that A. Artom had previously employed bent antennæ for radiotelegraphy, and also independent crossed straight antennæ with the object of creating circular and elliptically polarised electric waves.

CHAPTER VI

OSCILLATION DETECTORS

1. Classification of Oscillation Detectors. — In the previous chapters it has been shown that electric oscillations set up in an open or closed radiative circuit create an effect in the space around which is propagated through the dielectric as a wave, and that when these waves impinge in the right direction upon another open or closed syntonic receiving circuit placed at a distance, they set up in the latter similar oscillations, which, however, are far more feeble than the oscillations in the radiating circuit. These induced oscillations are not directly appreciable by our senses, and their existence can only be ascertained by the employment of special appliances, called *oscillation detectors*, inserted in or associated with the receiving circuit. The oscillation detector in its turn can be made to actuate some form of telegraphic or telephonic instrument, and thus to make evident to the human eye or ear the commencement, end, or continuance of the oscillations set up in the transmitting circuit.

In this chapter we shall consider the various forms of oscillation detector which have been invented. Of whatever type they may be, an oscillation detector is an appliance which enables us to detect the existence of a very small, high frequency alternating current in a circuit, or alternating difference of potential between two points on it. Hence, a first classification of oscillation detectors may be made by dividing them into *potential indicators* and *current indicators*. From this point of view they may be regarded as very sensitive forms of alternating current voltmeter or ammeter adapted for detecting or measuring exceedingly feeble but high frequency alternating electromotive forces or currents which exist in a circuit traversed by electrical oscillations.

A large number of forms of oscillation detectors have now been devised, depending upon the power of electric oscillations to effect various changes, as follows :—

1st. The simplest and the original method of detecting the

o

presence of oscillations in a circuit consists in the observation of spark discharges passing between points on the circuit, between which there is a difference of potential. Any arrangement for making such observations may be described as a *spark detector*.

2nd. The next in historical order is a method depending upon the power of electric oscillations when passed through a loose or imperfect contact between certain substances to make the electric conductivity of that contact better or worse. In the majority of cases the contact is improved and the conductivity increased, and since the surfaces are then more or less perfectly made to cohere together, such devices have been usually called *coherers*, but the term is not sufficiently general, and it is therefore better to speak of them as *imperfect contact* oscillation detectors.

3rd. Those depending on the power of electric oscillations to affect the magnetic properties or state of magnetic metals. These are called *magnetic detectors*.

4th. A large class which depend upon the ability of electric oscillations to heat a fine wire or substance of high resistance. These are called *thermal detectors*, or if a thermoelectric junction is employed, they are called *thermoelectric detectors*.

5th. A fifth type of oscillation detector operates in virtue of the power of electric oscillations to affect chemical action. These are called *electrolytic detectors*.

6th. A sixth type of detector depends for its action upon the unilateral conductivity of rarified gases, which are permeated by negative ions or corpuscles. These are called *valve detectors*, or *ionised gas* detectors.

7th. A type of detector has been invented based upon the possession by certain crystals, or the contact point between certain conductors, of the curious property of unilateral conductivity, or power of conducting a current much better one way through them than in the opposite direction. These are called *rectifying* detectors.

8th. Detectors based upon electrodynamic actions between fixed and movable conductors, one of which is traversed by oscillations. These are called *electrodynamic detectors*.

We shall briefly consider each of these classes of detector in turn.

2. **Spark Detectors.**—In the original researches by which Hertz experimentally established the production and properties of electromagnetic waves, he made use of a simple, closed, resonant receiving circuit having a small spark gap in it as a means of detecting electric oscillations set up in that circuit when held in various positions, in a space in which electromagnetic waves were falling upon it. This circuit was generally circular, and the

spark gap consisted of two small balls, adjustable by a micrometer screw (see Fig. 1a). If such a circuit is held with its plane perpendicular to the direction of the incident electromagnetic wave, and the line joining the spark balls of the resonator parallel to the direction of the electric component of the wave, small sparks are seen, due to the electric force in the wave setting up a potential difference between the spark balls. No such spark is seen if the resonator is held with the line joining its spark balls at right angles to the direction of the electric component of the wave.

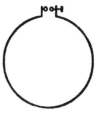

Fig. 1a.

The above arrangement constitutes a closed receiving circuit. An open one may be made by using two rods of adjustable length, which can be placed in one line, having adjustable spark balls between them (see Fig. 1b). If this open resonator is held

Fig. 1b.

parallel to the electric component of an incident wave, and if the lengths of the rods are adjusted to be about two-fifths of the wave length, then small sparks will be seen at the gap, due to the electromotive force set up in the rods by the incident wave.

This spark method of detection is, however, chiefly useful in laboratory experiments, and has a limited range of utility in radiotelegraphy on account of want of delicacy.

We need not, therefore, discuss the theory of the Hertzian resonator in this manual at greater length, but refer the reader to larger treatises for a fuller description of its properties and uses.

3. **Imperfect Contact Detectors.**—The first type of oscillation detector which was devised, having sensibility sufficient to be useful in radiotelegraphy, was developed out of the researches of numerous physicists on the peculiar conductive properties of metallic filings and loose contacts. As far back as 1835, Munk had observed that a mixture of tin filings, carbon, and other materials in a loose condition was non-conductive to electricity, but became conductive on passing the discharge from a Leyden jar through it. The same fact was observed again by Calzecchi-Onesti, in 1884. S. A. Varley, in 1852, noticed a remarkable fall in the resistance of masses of metallic filings under the action

of atmospheric electric discharges. In 1866, C. and S. A. Varley applied this discovery in the construction of a lightning protector for telegraphic instruments. In 1878, D. E. Hughes appears to have discovered that a tube full of zinc and silver filings placed in series with a voltaic cell and a telephone, became conductive under the action of an electric spark at a distance. In 1890, E. Branly, of Paris, rediscovered this important fact, and confirming the observations of previous researches, added much new knowledge. He noticed that an electric spark had the power of suddenly changing the electric conductivity of a loose mass of metallic filings placed a long way from the spark. In the majority of cases the change is from a poor conductivity to a much better one, but in a few cases, such as a loose contact between lead and peroxide of lead, the change is from a fairly good conductivity to a worse one. Branly constructed his metallic filings spark detector by placing in a tube of non-conducting material some metallic filings loosely packed between two metal plugs (see Fig. 2). He connected this appliance in series with

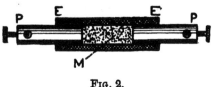

FIG. 2.

a galvanometer and a single cell, and by adjusting the pressure of the plugs was able to prevent the current from the voltaic cell affecting the galvanometer, because the loosely aggregated filings are non-conductive for the feeble electromotive force of a single cell. Under the influence of an electric spark at a distance, the filings, however, change quite suddenly into a condition of better conductivity, and the current from the cell passes through the mass and deflects the galvanometer. Branly found the same effect takes place when a loose or imperfect contact is formed between two pieces of slightly oxidised metal, such as copper or steel wires, and he found that this contact drops in resistance from many thousands of ohms to a few ohms under the influence of an electric spark made some yards away. He also noted that a slight tap or blow destroyed the improved contact.

These observations did not attract attention in England until described by Dawson Turner, in 1892, and they were then repeated by Croft before the Physical Society, in 1893, and carefully examined by Minchin and Lodge in the same year. In 1894, Lodge made use of a tube of glass full of loosely aggregated iron or brass filings contained between two plugs to detect the existence of electric waves created by a spark discharge. He christened this device a *coherer*, because he considered that the action of the oscillations produced by the incident electric wave was to make the particles cohere together.

A similar type of detector was employed by Popoff in Russia in researches on atmospheric electricity, and described by him at the beginning of 1896. In the same year G. Marconi described in a British Patent Specification a greatly improved form of metallic filings oscillation detector constructed in the following manner :— In a small glass tube about 3 or 4 cms. long and 5 mms. internal diameter, he placed two silver plugs fitting the tube tightly. To these plugs were attached platinum wires sealed through the closed ends of a tube. The inner ends of the plugs were polished and slightly amalgamated with mercury and brought within a couple of millimetres of each other. The interspace was filled with a very small quantity of nickel and silver filings, 95 per cent. nickel and 5 per cent. silver, carefully sifted. The glass tube was then exhausted and sealed. Subsequently the ends of the

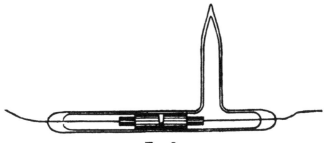

FIG. 3.

silver plugs were bevelled off so as to make the interspace wedge-shaped (see Fig. 3).

Marconi thus constructed an extremely sensitive form of imperfect contact oscillation detector, which under the influence of very feeble oscillations set up by electric waves passed from a condition of high resistance to a condition of low resistance. Branly had previously shown that a contact detector of this kind could be brought back to its original high resistance and sensitive condition by giving it a slight blow or tap, so as to wrench asunder the surfaces connected together by the action of the oscillations which had passed through it. Hence a metallic filings tube of this kind when employed for the detection of electric waves must be associated with some arrangement for continually tapping or decohering the filings. Lodge did this originally by means of a clockwork arrangement for continually shaking the metallic filings tube, or used an electric bell hammer to administer small blows continuously to the tube itself. The same was done by Popoff, in 1906, but he arranged the metallic filings tube in series with an electric relay, which, in turn, set in operation an electric bell, the hammer of which administered one or two blows to the coherer tube, shaking up the filings and bringing them back again to a non-conductive condition.

Marconi improved upon this by making all the adjustments capable of very nice regulation, and arranging the electromagnetic tapper so as to administer a delicate blow to the metallic filings tube from the under side of exactly the right strength to destroy the conductivity of the metallic filings immediately a current passes through them in virtue of the passage through the tube of electric oscillations (see Fig. 4).

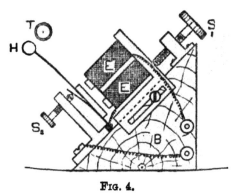

FIG. 4.

The final outcome of all this work was the invention of a complete apparatus, comprising a tube with metallic filings contained between two metallic plugs. This was placed in series with one or two voltaic cells and a telegraphic relay, so that when a small current passed through the metallic filings tube the moment it became conductive, this current was made to close another circuit and set in operation other telegraphic instruments. The relay was caused, in addition, to close a circuit through another electromagnet operating the tapper, so that the moment the metallic filings tube had passed into the conductive condition a blow was administered to it which brought it back again into the non-conductive state, whilst, at the same time, the relay was made to actuate some form of recording instrument, as more particularly described in Chapter VII.

The complication introduced by the necessity for tapping these metallic filings conductors back into a non-conductive condition led to inventions to construct self-restoring imperfect contact detectors. Branly had discovered that a loose contact formed with powdered peroxide of lead increased in resistance by the action of an electric spark. S. G. Brown has more recently constructed and described an auto-restoring oscillation detector, in which a small plug of dry compressed peroxide of lead is gently compressed between a platinum plate and a lead plate, but it is claimed for this device that it acts as an electrolytic valve, and will be, therefore, referred to again in a following section.

Lodge, Muirhead, and Robinson devised a form of contact detector consisting of a steel disc rotated by clockwork, the edge of which just touched the surface of some mercury covered with oil (see Fig. 5). The steel disc has a sharp knife-like edge, and under ordinary circumstances the film of oil carried round by the steel prevents good electric contact between the steel and the mercury ; but if electric oscillations are passed through the contact they break down the insulation of the film of oil and make a good electric contact between the steel and the mercury which, however, disappears as soon as the oscillations cease.

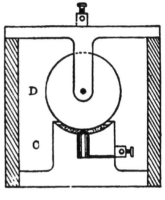

FIG. 5.

A form of self-restoring contact detector was devised by Italian naval officers, attributed by Captain Bonomo to Castelli, a signalman in the Italian Navy, but also claimed by the Marquis Solari. In this appliance a small globule of mercury is contained between a steel and carbon plug fitting tightly in a glass tube (see Fig. 6). A telephone and a single voltaic cell are included in series. By adjusting the pressure, the current from the cell can be prevented from flowing through the contact points between the mercury and the steel and carbon plugs, but if electric oscillations are passed

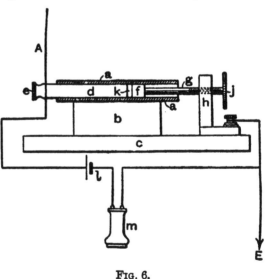

FIG. 6.

through these contacts the conductivity is improved and a current passes, but the conductivity automatically disappears on the cessation of the oscillations.

One of the best and simplest of these self-restoring imperfect contact oscillation detectors is the Tantalum-Mercury contact invented by L. H. Walter.

A fragment of tantalum wire from a tantalum lamp is attached to a platinum wire, and the tip of the tantalum immersed in mercury. The wire and mercury are connected respectively to a shunted voltaic cell in series with the telephone. This fraction of a volt of electromotive force is unable to send a current across the surface of contact of the tantalum and mercury, possibly because the tantalum is not wetted by mercury. When, however, oscillations are passed across the junction, the contact is improved so far that a current can pass through the telephone, creating a sound; but this contact between the tantalum and mercury ceases the moment the oscillations stop—hence the contact is self-restoring. The mercury with the tantalum point dipping into it can be sealed up air-tight in a glass vessel, so that the mercury is preserved from oxidation.

It will be seen, therefore, that all these forms of detector consist of some form of contact, either single or multiple, between various substances, the conductivity of this contact being either greatly improved or else diminished by passing electric oscillations through the contact point.

In the construction of imperfect contact detectors for radiotelegraphy the most important requirement is certainty of action. To secure this only a small testing current must be passed through the contact to work the relay or other indicating instrument which reveals the resistance change. The oxidisable metals are not so suitable for making these oscillation detectors as mixtures of rather oxidisable metals with others which are non-oxidisable. Hence, nickel and silver or aluminium and steel contacts are better than copper with copper. The magnetic metals nickel and iron, for some reason not fully understood, have a marked superiority over other metals. Hence, good contact detectors have been made by a light contact between the steel balls contained in a tube or, as in the tripod coherer of Branly, constructed by making a small copper three-legged stool, having the feet slightly oxidised, placed on a polished steel plate.

Many devices have been employed for restoring the contact to its sensitive condition, not only by tapping, but by rotating the tube or surfaces, or acting upon the metallic filings, if of iron or nickel, with a magnet. It is advisable, however, to exclude the air, so as to prevent the surfaces from becoming too much oxidised.

The metallic filings detector, in whatever form it may be made, must be regarded as a potential indicator, depending, as it does, upon the production of a certain small alternating difference of potential between two surfaces in loose or imperfect contact, the result of which is to improve or diminish the conductivity between these surfaces.

Much discussion has taken place upon the exact nature of the processes at work when electric oscillations pass through an imperfect contact. The original idea was that the effects were due to thermal action, heat being developed at the imperfect contact which welded together the junction. This hypothesis, however, fails to explain the contact action when such unweldable substances as carbon or other non-metallic conductors are employed. Also, it does not explain the decreased conductivity in the case of a lead and peroxide of lead junction. When the oscillations begin to take place across the junction there is a certain difference of potential between the points or surfaces in imperfect contact. We may think of the surfaces as initially separated by an extremely thin film of air and forming therefore a condenser of small capacity. Another view, therefore, taken of the effect is that the electrostatic attraction between these surfaces squeezes out the air and brings the surfaces into molecular contact, thus effecting an improved conductivity. At present, however, our knowledge of the true nature of electric conduction and of the reason some substances are better conductors than others is too imperfect to enable us to account for the fact that oscillations passing across a loose contact between two surfaces do not always cause an increase of conductivity. Neither the welding theory nor the electrostatic attraction theory explain why the magnetic metals, particularly nickel, are so much better than most others in making imperfect contact oscillation detectors. In spite of much research, therefore, the scientific problems in connection with this coherer action are by no means solved.

4. Magnetic Oscillation Detectors.—It was well known in the early part of last century that the discharge of a Leyden jar can magnetise steel sewing needles. Between 1842 and 1850, Joseph Henry, in the United States, examined this effect with great care, and only obtained a clue to numerous puzzling phenomena when he realised that the discharge of a condenser through a circuit of low resistance is oscillatory in character, and sometimes has the effect of magnetising and at other times of demagnetising the steel.

In 1870, Lord Rayleigh in discussing some electromagnetic phenomena pointed out that the effect of an oscillatory discharge of electricity upon iron or steel depends upon the direction of the maximum value of the current during the oscillations, and also that there may be superimposed magnetic effects in the needle. The subject was taken up again by E. Rutherford, in 1895, and in a Paper published in 1896 he described a number of important experiments in which the demagnetising power of electric oscillations was employed as an indicator for electric waves, making therefore the first magnetic detector, as follows: About twenty pieces of fine steel wire, each about 0·07 mm. in diameter and about 1 cm. in length, were insulated by shellac varnish and bound together into a little bundle. A fine copper wire insulated with silk was wound over the bundle in two layers of 80 turns. This small electromagnet was fixed at the end of a glass tube, and was placed at the back of a small suspended magnetic needle. If a current was passed through the coil it magnetised the steel wires to saturation, and they then caused a certain steady deflection of the suspended magnet. If, then, electric oscillations were sent through the coil wound round the steel wire, these demagnetised the steel, and caused the magnetometer deflection to diminish. Rutherford connected the two ends of the magnetising coil with two long horizontal rods acting as antennæ, and half a mile away he set up a Hertzian oscillator with its rods also in a horizontal position. After magnetising the steel wires the Hertzian oscillator at a distance was set in action. The electromagnetic wave passing out through space fell upon the receiving rods, and set up in them electric oscillations which, passing through the coil, demagnetised the steel wires, and therefore caused a diminution in the deflection of the associated magnetometer needle.

In 1897, E. Wilson took up the subject, and endeavoured to make this magnetic detector self-acting, so that the deflection of the magnetometer needle should close the circuit, and again magnetise the steel bundle. In 1902, G. Marconi described two other forms of magnetic detector. In one a thin bundle of iron wires was surrounded by a magnetising coil and its central part also embraced by a second coil in series with a telephone. Over this small bundle of iron wires a horseshoe magnet was slowly rotated so as to carry the iron through a cycle of magnetic changes, magnetising it first one way and then the other. If electric oscillations are sent through the coil wound round the iron they exercise an oscillatory magnetic effect on the iron, and these cause sudden magnetic changes in it which set up electromotive forces in the secondary coil embracing the iron, which make themselves evident by short sounds or ticks heard in the telephone.

Whether the change in the iron be from a less to a greater or greater to a less state of magnetisation, an inductive effect on the coil in series with the telephone is equally exerted, and the great sensitiveness of the telephone to sudden but small changes in the current through it makes it an extremely delicate means of detecting these small changes in the magnetic state of the iron.

For telegraphic purposes Marconi invented a far more perfect instrument, automatic in action, and capable of giving telegraphic signals, made as follows: Two wooden discs *e, e,* (see Fig. 7), grooved on the edges, are driven round slowly by clockwork. An endless band, *a, a,* made of a bundle of fine silk-covered iron wires, is arranged like a belt over these pulleys, and moves forward at the rate of 7 or 8 cms. per second. At one place the iron band passes through a glass tube, *g, b,* on which is wound a coil of insulated wire through which oscillations can be passed, and this coil is embraced in the centre by another coil, *c,* connected with the telephone, T. A pair of horseshoe magnets are placed with their similar poles together opposite to the last-mentioned coil, as shown in the diagram. If electric oscillations are passed through the coil wound round the band, they change the magnetic state of the iron, and generate an induced current in the secondary coil, and hence a current and a sound in the telephone. If oscillations continue to pass through the demagnetising coil at short intervals, the telephone will emit a sound which is

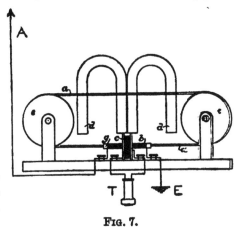

FIG. 7.

practically continuous, and if the oscillations are in longer or shorter groups, the corresponding sounds are produced in the telephone, and may be interpreted in accordance with an agreed code of signals. The extreme sensitiveness of the telephone to induced currents bestows upon this apparatus very great power in detecting feeble oscillations. The operation of this instrument was considered at one time to be due to the power of electric oscillations passing through the coil surrounding magnetised iron to annul the hysteresis of the iron. A reduction in magnetic hysteresis does not, however, invariably accompany the action of electric oscillations on iron or steel. Walter and Ewing discovered that in hard steel an increase in hysteresis may result when oscillations are sent through iron whilst being submitted to cyclical magnetisation. They devised a form of magnetic oscillation detector based on this fact as follows : An electromagnet rotates round a vertical axis so as to produce a revolving magnetic field (see Fig. 8). In the centre of that field is suspended a closed coil of hard drawn insulated steel wire suspended by an elastic wire. When the magnet rotates it tends to carry the suspended coil round with it in virtue of a dragging action due to magnetic hysteresis, and the torque so produced is resisted by the control of the elastic suspension. When electric oscillations are passed through the closed coil of steel it is found that the hysteresis of the metal is increased, and the coil twists more in the direction of the rotation of the magnet. This instrument, therefore, can be employed as a means of measuring the intensity of electric oscillations as well as merely to indicate their presence.

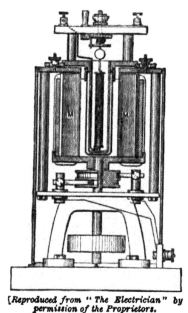

Fig. 8.

Another magnetic oscillation detector suitable for quantitative work has been devised by the author as follows :—

In a pasteboard tube are included 7 or 8 small bundles of fine iron wire, each wire well painted with shellac varnish, and each little bundle wound over uniformly with a magnetising coil of fine silk-covered wire, and over this coil and separated from it by guttapercha tissue another demagnetising coil of somewhat thicker insulated wire. The magnetising coils are all connected in series in such a manner that when a current passes through them, it magnetises the whole of the wires so that contiguous ends have the same magnetic polarity.

The outer or demagnetising coils are joined in parallel. Over this bobbin is wound a long secondary circuit consisting of 6000 turns of very fine silk-covered copper wire. Associated with this induction coil is a rotating commutator (see Fig. 9) carrying four insulating discs secured to a steel shaft. These discs have brass sectors let into their circumference, and against them four springs of brass wire press. As the commutator rotates, these springs and discs are made to close or open various electric circuits. The function of the first disc is to make a break in the circuit of the magnetising coils placed round the iron bundles, to magnetise them during one portion of its rotation, and leave them magnetised during the other portion. The function of the discs 2 and 3 is to short circuit the terminals of the secondary coil during the time the magnetising current is being applied by disc 1. A sensitive galvanometer is connected to the ends of this secondary coil, one being permanently connected and the other through the intermittent contact made by disc 4. The operations which go on during one complete revolution of the discs are as follows : First the current of a battery of secondary cells is employed to magnetise the iron bundles, and during this time the terminals of the fine wire secondary coil are short circuited, and the galvanometer disconnected. Shortly after the magnetising current is interrupted, the secondary bobbin is unshort circuited, and immediately afterwards the galvanometer circuit is completed. Hence, during a large part of one revolution the iron wire bundles are left magnetised and surrounded by a secondary coil connected to a galvanometer. If during this period electric oscillations pass through the demagnetising coils, an electromotive force is induced in the secondary bobbin by the

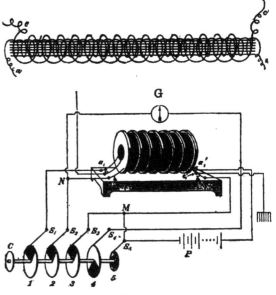

Fig. 9.

demagnetisation of the iron, and this causes a deflection of the galvanometer coil. Since the interrupter discs are rotated very rapidly, if the oscillations continue, this intermittent electromotive force produces a practically steady current through the galvanometer which is proportional to the demagnetising force being applied to the iron. Hence the arrangement becomes not merely a means of detecting oscillations, but of measuring their intensity. This instrument has been employed by Buscemi for quantitative measurement of the opacity of various dielectrics to electric waves.

The endeavour to account for these interesting magnetic effects of electric oscillation has given rise to much research and discussion. The chief contributions to it have come from Maurain, L. H. Walter, Ascoli, Arno, Piola, Foley, C. Tissot, P. Duhem, W. H. Eccles, and J. Russell.

Russell has carefully distinguished between the two conditions under which we can work.

(i.) Iron or steel may be placed in a constant magnetic field, and then subjected to the action of electric oscillations.

(ii.) Iron or steel may be subjected to continuing electric oscillations, and then the magnetic field around it changed.

In the case of Rutherford's experiment, hard iron or steel having considerable retentivity is subjected to a magnetic force, which is then removed, leaving remanent magnetisation in the iron. The action of oscillations taking place round the iron is then always to remove or diminish this magnetisation, and this can be detected by a change in the position of a suspended magnetic needle in the neighbourhood.

In Marconi's first form of magnetic detector, a horseshoe magnet is rotated slowly (about one turn in two seconds) over a thin bundle of hard iron wires, which are surrounded by two separate coils of wire. The iron is thus carried slowly through a cycle of magnetising force of equal positive and negative values. The magnetisation induced in the iron lags behind the magnetising force in virtue of so-called hysteresis, and, therefore, if ordinates representing the magnetisation are plotted out in terms of the magnetising force as abscissæ, we obtain a magnetisation curve of the well-known looped form. The area of this loop is proportional to the work expended in carrying the iron through one complete magnetic cycle. Russell states, as the result of his experiments, that if oscillations act continuously on the iron whilst it is being carried round the magnetic cycle, the area of this loop is greatly increased, thus showing an increase in hysteresis loss. If the oscillations come intermittently, as they would do in radiotelegraphic signalling, then the effect depends upon the particular point in the cycle at which the oscillations arrive. The result, in any case, is to produce a sudden change in the magnetisation of the iron. Hence, if the oscillations are sent through one coil wound round the iron, the sudden change in magnetisation produced by them creates induced or secondary electric currents in another coil wound over the oscillation coil, and therefore causes a sound in a telephone in series with the latter coil.

In Marconi's iron band form of magnetic detector the action is somewhat different. The band is passing through a magnetic field, so as to be always subject to a longitudinal magnetising force, which is first in one direction and then is quickly reversed, because the two horseshoe magnets are placed with their poles against the wire and have similar poles in contact. Under these conditions, Russell found that the effect of a longitudinal oscillatory magnetic force is to increase the magnetisation due to the steady force by an amount which is greater for an increasing than for a decreasing field. If the double north poles of the magnets are in the centre, and the iron moves from left to right, then the moving iron band distorts the field, and the effect of the oscillations passing round the iron is to increase the magnetisation of the iron more on the left hand than on the right of the north poles. Both these effects alter the number of lines of magnetic flux through the secondary coil in series with the telephone, and therefore cause an induced current to flow through it, and the telephone in series emits a sound. Hence, Russell considers that Marconi's second form of magnetic detector acts in virtue of the increase of magnetisation in iron which occurs when an oscillatory field is superimposed upon a slowly changing or stationary field near a cyclic extreme, whereas Rutherford's form of detector operates in virtue of a decrease of magnetisation produced when the magnetising force has been applied and has been removed.

The function of the moving band is twofold: it supplies the hard iron or steel in a condition of low permeability to be raised by the oscillations to a condition of higher permeability, and it distorts the field in the direction of motion. The action of this moving iron band form of detector cannot be explained by describing to the electric oscillations merely the power of effecting a reduction of magnetic hysteresis.

According to W. H. Eccles, the effect produced when a bundle of iron wires is taken slowly round a magnetisation cycle and an oscillatory magnetising force applied at any point is to bring the iron back to the condition of magnetisation it would have under the final steady impressed magnetic force acting on it if the hysteresis was suddenly annulled. The action of the oscillations is therefore to cause a return to the normal curve of magnetisation.

In the Walter-Ewing form of detector we have different magnetic conditions. The field is then revolving somewhat rapidly, so that a drag is produced on the suspended iron, due to the so-called rotational hysteresis. Oscillations increase this hysteresis, and therefore the deflection of the suspended iron at least in fairly strong fields.

According to L. H. Walter, all magnetic detectors may be divided into two classes. First, those in which the oscillations act on the iron after the magnetising force has been applied and withdrawn. Examples of this type are the original Rutherford detector and the Fleming quantitative detector above described. In this case the available energy is limited to the remanent magnetism in the core, and the action of the oscillations is to reduce or destroy this remanent magnetism. The second class, represented by Marconi's moving band detector, derive their energy from an external magnetic field and from the motive power driving the band, and the action of the oscillations is merely to release some of this energy. If the iron is moving through a field of increasing magnetic force, it is on the lower side of the hysteresis loop, and the action of the superimposed oscillatory field is to increase the magnetisation when not actually at the peak of the curve, the increasing effect being, as E. Wilson first showed, greatest at or near the point of inflection of the lower branch of the hysteresis curve. This increase in magnetisation of the portion of the iron band partly enclosed by the coil in series with the telephone creates the induced current in the latter. We may, therefore, in a sense, speak of this increase in local magnetisation of the iron as due to an annulment of hysteresis. The action of the moving band detector is, however, essentially dependent on the supply of energy from an external source to magnetise the iron and move it against the magnetic force. The action of the oscillations is only a trigger action, which creates a sudden increase in the magnetic flux in a part of the iron embraced by the secondary or telephone coil. This causes in turn induced currents to flow through it, first one way and then the other. Accordingly, with the band detector, the only possible signal receiving instrument is a telephone, unless we provide some means of sifting out the direct from the inverse induced current in the telephone coil. This has been achieved, however, by means of one of the author's oscillation valves or glow-lamp detectors, described in a subsequent section, and by its use it is possible to obtain from a Marconi moving band magnetic detector, associated with a Fleming oscillation valve, intermittent but unidirectional currents, which can operate a relay, and therefore work any ordinary telegraphic printing instrument.

Another method has been devised by L. H. Walter, by which a detector of the Walter-Ewing type, depending upon rotational hysteresis, can be made to furnish continuous currents. In this case oscillations are made to act on a magnetic mass undergoing reversals of magnetism in a rotating field in such a manner that the changes of magnetism produced by the oscillations create alternating induced currents in embracing coils of wire, which are rectified by a commutator in the usual dynamo machine manner. The inventor has described his apparatus as follows :—

Two ebonite bobbins, B, B (see Fig. 10), mounted on the same spindle, are rotated in the field of two horseshoe permanent magnets, NS, NS, these bobbins being wound, in a similar manner to those illustrated in connection with the pivoted bobbin detector previously referred to, with some feet of steel wire of suitable resistance. A wind-ing of two coils, W, W', at right angles to one another, of a hundred turns, is placed on each bobbin at right angles to the plane of the steel wire winding, as in a drum armature, corresponding coils, *i.e.* W and W, W' and W', being connected in such a way that the E.M.F.'s generated are equal and opposite. The ends of the windings are connected to the segments of a four-part commutator, C. (For the sake of clearness, only one pair of corre-sponding windings, of one turn each, is shown connected in Fig. 10.) The steel wire windings of the two bobbins are exactly alike, the ends of one winding being insulated, while those of the other are connected to a pair of slip-rings, *r, r,* and brushes, by means of which the oscilla-tions can be passed through the winding.

On testing this apparatus, with no oscillations acting, there was no potential difference at the brushes. On waves arriving, a steady deflection of the gal-vanometer was obtained in a direction corresponding to an increase of E.M.F. generated by the armature acted upon by the oscillations. By suitably proportion-ing the turns in the winding the sensibility was considerably increased. The usual speed employed is about five to eight

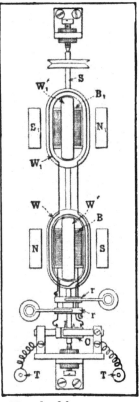

[Reproduced from " The Elec-trician" by permission of the Proprietors.

Fig. 10.

revolutions per second. Higher speeds have been tried, and give a larger effect, but the zero is not so steady. Telephonic signals can, of course, be received simultaneously by connecting to the winding at some point before the E.M.F. is commutated. When a relay alone has to be actuated, however, it may be advantageous to so arrange matters that the generated E.M.F.'s do not exactly

P

balance, and a small initial current, insufficient to actuate the relay, passes all the time through it. The change can be rapidly effected by a very slight shift of the brushes.

5. **Thermal and Thermoelectric Detectors.**—An electric oscillation, being a form of electric current, produces heat in a circuit through which it flows. The heat produced per second is proportional to the resistance of the conductor and to the square of the effective or R.M.S. value of the current. This mean square value is called the *integral value* of the current. Accordingly, if the oscillations consist of separate trains, these not being very close to one another, the integral value of the oscillations may be small, although the maximum value in each train may be very considerable. If we call J the integral value of the current —J being the R.M.S. value—and if we call I the maximum or initial value of the current in each train, δ the decrement of a semi-period, n the frequency of the oscillations, N the number of trains per second, then I and J are related as follows:

$$J = I \sqrt{\frac{N}{8n\delta}}$$

provided that the oscillations are not very strongly damped, as is the case with a closed oscillation circuit. If the oscillations are persistent or undamped, then if I is the maximum value, we have

$$J = I\frac{1}{\sqrt{2}}$$

provided the oscillations are of simple sine form.

To detect feeble oscillations by thermal effects, we must pass them through a wire of high resistance but small heat capacity, so that there may be a relatively large rise of temperature, and we must provide some means of detecting this rise of temperature in the wire. Since the introduction of a resistance in an oscillation circuit damps the oscillations, it is necessary to make the resistance introduced as small as possible and to localise the heat produced in it. Hence, what is used is a short length of not more than a few millimetres, or even less, of extremely fine wire made of a material of high resistivity, such as constantan or platinoid, and we must provide some sensitive means for determining a small rise of temperature in the wire. Since the rise of temperature measured depends upon the relative rate at which the wire gains and loses heat, we can increase the rise of temperature for a given heat production in the wire by reducing the convection of heat by the air. In other words, if we put the wire in a high vacuum we can obtain the greatest rise of temperature for a given integral value of the oscillations. Such an arrangement of a short length of high resistance wire enclosed in a vessel which may or may not be exhausted of its air, is called a *bolometer*, or, by Fessenden, a *barretter*. The small rise of temperature may be detected in several ways.

(1) We may enclose the wire in the glass bulb of an air thermometer, and make a form of electric thermometer, which was devised in the last century by Snow Harris, but by continental writers generally attributed to Reiss. To render it insensitive to changes of external temperature, this thermometer should be of a differential form, consisting of two glass bulbs connected to a glass U-tube, containing some non-volatile liquid, such as sulphuric acid. One of the bulbs has a fine wire sealed through it. The level of the liquid is not then affected by external changes of temperature, but if oscillations are passed through the fine wire, heat is produced, the air in that bulb is expanded, and the level of the liquid falls in one tube and rises in the other until a stationary condition is reached.

Such an instrument may be calibrated for integral values of the oscillations by passing various continuous currents through the wire and noting the position of the liquid in the U-tube. This instrument is generally called a hot wire thermometer.

(2) We may detect a rise of temperature by the change in resistance produced in the bolometer wire itself. This was first done by Rubens and Ritter in 1890, who constructed a form of Wheatstone's bridge, of which two of the arms consisted of rectangles of fine wire, the other two arms of variable resistance as usual, the galvanometer G and battery B being connected, as shown in Fig. 11. If, then, electric oscillations are passed through one of the rectangles, R, from corner to corner, this arm of the bridge is heated, its resistance is increased, and the balance of the bridge upset, the galvanometer G showing a deflection, which indicates a change in resistance, and therefore serves to measure the rise of temperature produced in the wire, and therefore the integral value of the oscillations.

C. Tissot has employed this method in a more refined form as a detector of oscillations produced by the electric waves utilized in radiotelegraphy. He makes use of a fine platinum wire not more than 0·01 mm. in diameter as the bolometer wire to be heated by the oscillations, and detects and measures this heat by the variation in resistance of the wire. With such an arrangement he states that he has detected electric waves at a distance of 40 kilometres from the radiator.

R. A. Fessenden has devoted much attention to the improvement
of thermal oscillation detectors, made as follows : An extremely
fine platinum wire, about 0·08 mm. in diameter, is embedded in
the axis of a silver wire, about 2 mm. in diameter, like the wick
of a candle. The compound wire is then drawn down until the
diameter of the platinum wire is reduced to about 0·0015 mm.,
in the manner first suggested by Wollaston. A short piece of
this Wollaston wire is bent into a loop and the ends attached to
stouter wires and enclosed in a glass bulb. If the tip of the
loop is dipped into nitric acid, the silver coating is dissolved off,
leaving a short length of exquisitely fine platinum wire sealed

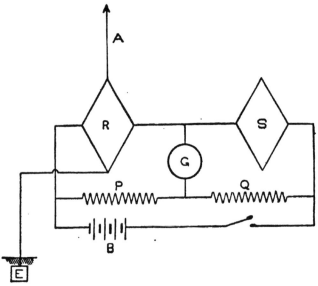

FIG. 11.

into the bulb like the filament of an incandescent lamp (see
Fig. ⁕12). The bulb is then closed and the air exhausted.
Furthermore, the glass bulb may be enclosed in a silver bulb to
shield it from external radiation. To detect the small rise in
temperature produced in this fine wire by very feeble oscillations passed through it, the ends of the loop are connected to
a telephone, a single shunted cell being interposed. The cell
then transmits a feeble continuous current through the telephone
and the bolometer wire. If oscillations are then sent through
the latter, they still further heat the loop and suddenly decrease
the current through the telephone. The ear then detects this
decrease of telephonic current by a sound made in the telephone.

Fessenden found that in place of the platinum wire a fine tube filled with a high resistance liquid could be employed, or even the two ends of a very fine platinum wire dipped into a liquid, thus forming what he calls a liquid barretter. A number of such bolometer wires can be arranged in parallel to be heated by the same oscillations.

(3) We may use a thermoelectric couple to detect a small rise of temperature in the wire. Such an arrangement was first employed by Klemencic. The oscillations are sent through a fine constantan wire, and against this rests the junction of a thermoelectric couple made of iron and constantan, or some other pair of suitable metals. The ends of the couple are connected

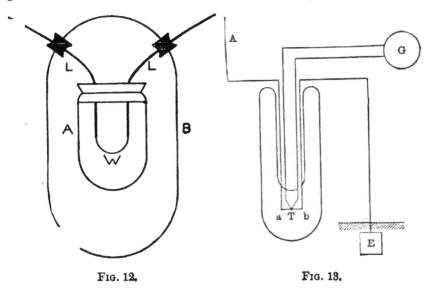

Fig. 12.   Fig. 13.

to a suitable low resistance galvanometer, and on passing oscillations through the high resistance wire the galvanometer gives an indication.

An improved arrangement of this kind was devised by the author in 1906, taking advantage of the position of tellurium and bismuth in the thermoelectric series. A sort of double test-tube of glass was constructed (see Fig. 13), the interspace between the two tubes being subsequently exhausted. Through the bottom of the inner test tube were sealed four wires, two of these, *a*, *b*, were connected to a fine constantan wire, and the other two, *c*, *d*, were connected to a tellurium-bismuth thermojunction T formed of very fine wires of bismuth and tellurium, the junction being soldered by a special solder to the centre of the constantan wire. A high vacuum was then made in the interior space. When oscillations are passed through the constantan wire, a suitable low resistance galvanometer being connected to the leads from the thermojunction, the galvanometer deflects, the deflection being proportional to the square of the integral value of the oscillations. This detector has proved to be of great use in quantitative researches.

Another form of thermoelectric detector was devised by Mr. Duddell. He employs a form of Boys' micro-radiometer in which

a thermocouple of bismuth and antimony wire forming a loop is suspended by a quartz fibre, in a strong magnetic field. A mirror attached to the thermocouple enables small deflections to be measured. Underneath this thermocouple he placed a very fine narrow strip of metal, say gold leaf, through which the electric oscillations are passed. These heat the strips feebly. One junction of the small suspended thermocouple rests just above the strip, but not quite touching it, and is therefore heated by radiation and convection. The couple is therefore traversed by a current and deflected in the magnetic field. This

[Reproduced by permission of the Cambridge Scientific Instrument Company.

Fig. 14.

thermal detector is also quantitative. The complete instrument is shown in Fig. 14.

(4) We may detect the heating of the bolometer wire by its own expansion, as first done in 1889 by W. G. Gregory. This arrangement is more suitable for detecting large and powerful oscillations than feeble ones. The author has devised and used a hot wire ammeter made as shown in Fig. 1, Chapter VIII., where the oscillations are passed through a fine wire, the ends of which are fixed. The expansion of the wire therefore produces a sag of the middle part, which is magnified by an index needle or by the movement of a mirror and a ray of light. The advantage of enclosing the bolometer in a vacuum is very considerable, because in the case of fine wires heated by a current the removal

of the heat is chiefly effected by air convection, and if therefore we stop this convection by removing the air, the temperature of the wire rises higher for a given integral effect of the oscillations.

These thermal detectors are very useful for quantitative work in the laboratory, because they can be so easily calibrated by passing through them a continuous current, and they are therefore not merely oscillation detectors but are measurers of the integral value of the oscillations or of their mean-square value. Moreover, the thermal detectors are especially valuable in connection with the measurement of undamped oscillations, as there we have to do with integral values which are large compared with the integral values of damped oscillations.

Another form of thermal detector can be made by dispensing with the bolometer wire and heating the thermoelectric junction itself directly. Thus, for instance, a form of detector has been made by L. W. Austin which consists of a fragment of tellurium pressed against the edge of an aluminium disc, the pressure being adjustable by a screw. A telephone is connected across this junction. If oscillations are sent through the junction they produce heat at the point of imperfect contact and therefore a thermoelectric current, which passes through the telephone and will therefore give an indication, if the oscillations are started or stopped, by a sound in the telephone. Such an arrangement, therefore, is not suitable for metrical work, but is suitable as a radiotelegraphic detector where trains of oscillations are intermittently created and stopped.

6. Electrolytic Oscillation Detectors. — Numerous isolated observations were made between 1898 and 1902 which indicated that electric oscillations possess the power to affect the polarisation of small metallic surfaces immersed in an electrolyte.

In 1898, A. Neugschwender showed that if a deposit of silver was made on a sheet of glass and divided into two parts by a sharp cut with a razor, and a film of moisture deposited on the glass, then electric oscillations taking place across the gap had the power to alter the resistance of the film of moisture.

Similar observations were subsequently made by E. Aschkinass and L. de Forest, and the latter made an oscillation detector consisting of a tube closed at the ends with metallic plugs, the interspace being filled with a mass of peroxide of lead and glycerine having in it metallic filings.

The modern form of electrolytic detector originated in 1903 with R. A. Fessenden, and was shortly afterwards independently invented by Schlömilch. It consists essentially of a vessel having as one electrode a very fine short wire of platinum, offering therefore an extremely small surface. This is generally made the anode or positive pole. The other electrode is a platinum or lead or silver plate of much larger surface, and the two are immersed in an electrolyte, which may be nitric acid, dilute sulphuric acid, or any other aqueous electrolyte yielding oxygen or hydrogen on electrolysis (see Fig. 15). The electrode of small surface is generally prepared from a Wollaston wire by drawing down a platinum wire coated with silver, or else a platinum wire coated with iron, until the platinum wire itself is less than 0·001 mm. in diameter. A short length of the compound wire is then fixed to the end of a screw, so that it can be lowered by a very small amount into the electrolyte. If, for instance, the electrolyte is nitric acid, then a silver-coated wire is employed, and on lowering the tip of this into

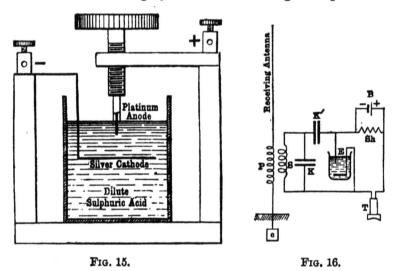

FIG. 15.   FIG. 16.

the acid the silver is dissolved away, leaving a platinum electrode of microscopic dimensions immersed in the liquid. If dilute sulphuric acid is employed, then an iron-coated platinum wire is used. If a small electromotive force is applied to such a cell by means of a shunted voltaic cell, B, and a telephone, T, or sensitive galvanometer included in the circuit, a small current will flow through the electrolytic cell, E, and will polarise the electrodes, which will thereupon reduce the current practically to zero (see Fig. 16). Under these conditions, if oscillations are sent through the electrolytic cell they destroy the polarisation of the small electrode and the current suddenly increases, but it returns to its former small value as soon as the oscillations cease.

According to Schlömilch and Lee de Forest, this electrolytic detector is sensitive to oscillations only when the small electrode is connected to the positive pole of the voltaic cell ; that is, when it is polarised with oxygen gas, but Fessenden, Rothmund, and Lessing state that the cell is equally sensitive when the point is negative. According to L. W. Austin, for feeble oscillations the cell appears to be about equally sensitive both ways, but for stronger depolarising currents, acts best with the small electrode positive. There is likewise a difference of opinion as to the reason for the operation of this detector. Fessenden advocates a thermal theory, according to which the chief effect is due to the change of resistance produced at the surface of the small electrode, but J. E. Ives has found that if platinum black is deposited on the small platinum electrode this deposit, as is well known, reduces the polarisation effect and stops the detector action. If the cell is used with a telephone in series with a shunted voltaic cell, and if one of the electrodes in the electrolytic cell is connected with an antenna, and the other with the earth or a balancing capacity, then the impact of electric waves upon the antenna will cause oscillations to pass through the electrolyte and the sudden increase in the current through the telephone is heard as a short sound or tick. If the trains of oscillations are intermittent but rapidly succeed each other, then these short sounds in the telephone run together to a continuous noise, and in this manner by acting upon the cell by a series of trains of oscillations more or less prolonged, signals according to the Morse alphabet can be conveyed. The valuable feature of this electrolytic detector is that it is not merely qualitative but quantitative. The degree to which the polarisation is destroyed is in some sense proportional to the amplitude of the oscillations. Hence, when employing the instrument with a galvanometer, the deflection of the galvanometer will vary according to the amplitude of the oscillations passing through the cell. It is also sensitive to a wide range of frequency and can be employed with alternating currents of low frequency, and under these circumstances can detect a voltage of a few ten-thousandths of a volt applied to the terminals of the electrolytic cell. The resistance of the electrolytic cell with slowly alternating currents may vary from a few hundred to many thousand ohms. Hence from one point of view we may regard the electrolytic oscillation detector as a detector of the same type as the imperfect contact detector, in that the action of oscillations is to produce an effect on it equivalent to a sudden decrease in resistance. If acted upon by undamped oscillations, the amplitude of which vary continuously, then the equivalent resistance of the cell varies continuously, and in some degree proportionately to the intensity of the electric oscillations.

This important fact has been applied, as shown in Chapter IX., in connection with radiotelephony, and has been made the means by which undamped electromagnetic waves are made to convey articulate speech between two points without the use of continuous connecting wires.

A form of detector which is by some classified as an imperfect contact and by others as an electrolytic detector is that invented by S. G. Brown. It consists of a pellet of peroxide of lead held between a plate of lead and one of platinum. If an external E.M.F. from a single secondary cell is impressed upon it so that the current flows through the peroxide from platinum to lead, this current will experience a counter electromotive force due to the electro-chemical action of the lead-peroxide of lead-platinum

[*Reproduced by permission of the Cambridge Scientific Instrument Company.*

FIG. 17.

couple. According to Mr. Brown, when oscillations pass through this couple they increase its counter-electromotive, force by stimulating chemical action and so reduce the current sent through it by the external cell. The couple acts, therefore, as a conductor of which the resistance is increased by electric oscillations. The pellet of peroxide is mounted up in a holder so as to apply to it an adjustable pressure (see Fig. 17), and is placed in series with a galvanometer and single cell. When oscillations are created through the peroxide the deflection of the galvanometer decreases but increases again when they cease.

7. **Valve or Rectifier Oscillation Detectors.**—Since electric oscillations are alternating currents of high frequency, the means of detecting them simply as electric currents which

have been already described are based upon actions which are independent of the direction of the current. Thus, for instance, the heating effect of an oscillation being determined by the square of the strength of the current at any instant, is independent of its sign or direction, and the same is true of the coherer or anticoherer action, as it is sometimes called, viz. the increase or decrease of the electric conductivity of an imperfect contact.

Speaking generally, the methods available for measuring an alternating current are vastly inferior in sensibility to our means of measuring a direct or continuous current. Thus no form of alternating current ammeter has yet been devised which is at all comparable in sensibility with the ordinary direct current mirror galvanometer. An instrument of the latter class can quite easily be made to produce a very large deflection of a spot of light upon a screen due to passage through the coils of the galvanometer of a current of one hundred millionth of an ampere. But it is extremely difficult to make an alternating current ammeter which will give any large deflection of a spot of light across a scale for a current of much less than one-thousandth of an ampere. Accordingly, we are much limited in our ability to detect electric oscillations by the fact that they are alternating currents. If, however, means are available for rectifying these high frequency alternating currents, that is to say, eliminating the movement of electricity in one direction, and converting them into continuous currents, then all ordinary sensitive mirror galvanometers become available as instrumental means for rendering visible or evident the presence of oscillations in the circuit. Hence any appliance for rectifying electric oscillations is an important addition to our means of detecting them. To achieve this it must possess unilateral conductivity, in other words, it must permit the passage of electricity through it in one direction but not in the other, or permit it to flow in one direction under much less electromotive force than in the other direction.

A very simple but effective form of oscillation valve was invented by the author in 1904, based upon researches made many years previously, viz. in 1890, upon the Edison effect in incandescent electric lamps. An ordinary incandescent lamp with carbon filament has a metal plate included in the glass bulb, or a metal cylinder, C, placed round the filament, the said plate or cylinder being attached to an independent insulated platinum wire, T, sealed through the glass (see Fig. 18). When the carbon is rendered incandescent by electric current, the space between the filament and the plate, occupied by highly rarefied gas, possesses a unilateral conductivity, and negative electricity will pass from the incandescent filament to the plate, but not in the opposite direction. This effect depends upon the now well-known fact that carbon in a state of high incandescence liberates electrons or negative ions; that is to say, point charges of negative electricity. These electrons, or corpuscles, are constituents of the chemical atom. Hence, a carbon filament in an incandescent lamp is discharging from its surface negative electricity, which may even amount to as much as an ampere or even several amperes per square centimetre. If, then, an incandescent lamp made as described has its filament rendered incandescent by a continuous current, and if another circuit is formed outside the lamp connecting the negative terminal of the filament with the insulated metal plate or cylinder in the bulb, and if oscillations are set up in this circuit,

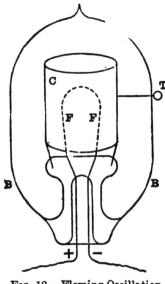

FIG. 18.—Fleming Oscillation Valve.

negative electricity will be able to move through this circuit from the filament to the plate inside the bulb, but not in the opposite direction. Hence, if an ordinary continuous current galvanometer is included in the external part of this last-named circuit, it will give a deflection when oscillations are set up in that circuit. For this purpose the author employs a small incandescent lamp with a rather thick carbon filament, or better still, a double or treble filament, taking altogether about 2 amperes at a terminal voltage of 12 volts. The filament must be so constructed that it is brightly incandescent when

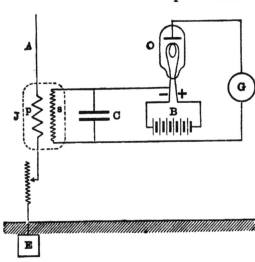

FIG. 19.

1.2 volts are applied to the terminals of the lamp, or, to use a lamp manufacturer's phrase, the filament must be working at an efficiency of at least 3 watts per candle. Another coil in which oscillations are being set up is then connected with the negative terminal of the filament and the external terminal of the metal cylinder. This circuit also includes a sensitive galvanometer, relay, or telephone. On setting up oscillations in that circuit they will be rectified, and the galvanometer will give a steady deflection as long as the oscillations last.

In using this oscillation valve or glow lamp detector as a receiving arrangement in radiotelegraphy, the author places the valve, O, in series with the relay or galvanometer, G, in the secondary circuits of an oscillation transformer, the primary circuit, *p*, of which is inserted in between the receiving antenna and the earth (see Fig. 19). The valve then rectifies the oscillations produced in the circuit, *s*, and the relay or galvanometer is affected.

Marconi has employed this glow lamp detector with a telephone in a slightly different manner, as shown in Fig. 20. The antenna, A, is coupled through an oscillation transformer with a circuit which includes the valve, O, and the fine wire coil of an ordinary 10-inch spark induction coil, I, the low resistance coil of which is in circuit with the telephone, T.

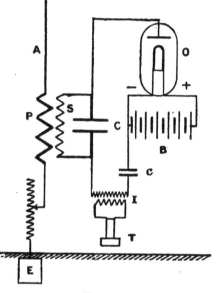

FIG. 20.

Condensers are placed across the secondary circuit of the oscillation transformer and also in series with the fine wire coil of the large induction coil. By suitable adjustments of the capacity of this condenser the circuits are brought into resonance. Oscillations taking place in the antenna, due to the impact of electric waves upon them, are then transformed by the oscillation transformer, rectified by the oscillation valve, and sent through the fine wire coil of the large induction coil in the form of unidirectional but intermittent currents, and these oscillations are again transformed up in current value by the large induction coil, and create a sound in the telephone. So used, the oscillation valve becomes one of the best long-distance receivers for electric waves yet devised.

It is a characteristic property of ionised gases that the current through them does not increase proportionately to the electromotive force, but soon reaches a value called the saturation current, beyond which no further increase takes place with increasing electromotive force. In the case of the author's glow lamp detector the current read on a galvanometer in series with the valve does not increase in value proportionately to the electromotive force of the oscillations rectified beyond a certain limit.

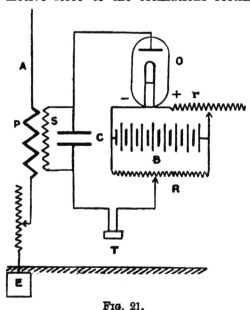

FIG. 21.

At or about 20 volts potential difference between the cylinder and negative terminal of the carbon filament valve there is a very marked rise in the current for a small increase of voltage. It is possible, however, to calibrate the arrangement of valve and galvanometer, so as to enable the integral value of the oscillations to be obtained from the galvanometer deflections. This glow lamp detector has been much used by Lee de Forest, disguised under the name of an *audion,* and claimed as his own invention. It was, however, described in scientific papers and in patent specifications by the author long previously.

The author has since found that greatly improved results can be obtained by employing a particular type of glow lamp with a Tungsten filament and a copper insulated cylinder surrounding it. The electronic emission from the Tungsten is greater than that from carbon, probably because it is a better conductor and can be raised without volatilisation to a much higher temperature than carbon. The author now uses this Tungsten glow lamp detector as follows : The battery which supplies current to incandesce the filament has a variable resistance, r, placed in series with it, and a high resistance, R, as a shunt across its terminals. A movable contact on this shunt, R, is connected through a telephone, T, with one plate of the condenser in the oscillation circuit of the receiving antenna (see Fig. 21), and the metal cylinder is connected to the other plate of the condenser. The two resistances can be so adjusted that the action of the oscillations taking place across the vacuous space in the valve is to increase the electronic emission, and therefore the currents through the telephone. If, therefore, trains of waves are infringing on the antenna, they are heard on the telephone as long or short sounds, depending on the number of trains incident.

Another class of valve or rectifier detector is based upon an interesting property possessed by certain crystals of rectifying electric oscillations. It was discovered by General H. H. C. Dunwoody, in the United States, that a crystalline mass of carborundum, which is an artificial silicate of carbon, when supplied with electrodes, acts as an oscillation detector, and converts these oscillations into a continuous current. The crystal is inserted in the circuit of the antenna, and shunted by another circuit containing a telephone and a battery. When oscillations are set up in the antenna sounds are heard in the telephone, and it was found that the battery may be dispensed with and yet sounds continue to be heard. This phenomenon has been carefully investigated by G. W. Pierce, and he found that the current through the crystal in one direction under a given electromotive force was very much greater than with a current of the opposite direction under the same electromotive force. In other words, the carborundum possesses a marked unilateral conductivity. This property had, moreover, been previously found in some crystal metallic oxides and sulphides by F. Braun, but none of these showed such striking asymmetry as that shown by carborundum. Thus, for instance, if a continually increasing voltage is applied to a mass of carborundum crystals, the following table, taken from a paper by G. W. Pierce, shows the currents produced in one direction or the other in microamperes by that voltage. It will be seen that under an impressed electromotive force of 10 volts, the current in one direction is a hundred times greater than in the opposite direction, but this ratio decreases with the rise of voltage.

RELATION OF CURRENT TO VOLTAGE, SHOWING UNILATERAL
CONDUCTIVITY OF CARBORUNDUM.

	Current in microampères.		
Volts.	C Commutator left.	C' Commutator right.	$\dfrac{C}{C'}$
10·0	100	1	100
12·1	150	—	—
12·8	200	—	—
14·5	300	5	60
16·0	400	—	—
16·8	500	10	50
17·7	600	—	—
19·4	700	—	—
20·0	800	20	40
21·0	900	—	—
21·9	1000	30	33
23·2	1200	50	24
25·0	1500	—	—
27·5	2000	120	17

Pierce has discovered that various other crystals possess the
same unilateral conductivity, and may therefore be used in the
same manner as oscillation detectors, when placed in the oscillation circuit and shunted by a telephone and battery. One of these
is the material called hessite, which occurs in nature as a telluride
of silver or gold; and it has also been found that a crystal of
oxide of titanium acts in the same manner. This crystal occurs
naturally as a mineral known as octahedrite or anatase. The
cause of this unilateral conductivity has been much discussed.
It is possible that the cause may be thermoelectric, but sub-
merging the crystal in oil does not appreciably change its
behaviour, nor heating one junction more than the other.
Pierce measured the current voltage or characteristic curve
of carborundum, considered as a conductor, and found that
it was not linear, but that the apparent resistance of the substance falls as the current is increased, which implies a decrease
of terminal potential difference with increase of current. In
this respect, therefore, the carborundum and the other crystals
resemble the electric arc and ionised air rather than metallic
conductors.

This, however, is not a complete explanation, because, as Dunwoody points out, carborundum may be used as a detector of
electric waves without any battery in the circuit. Nevertheless,

the property of the material above mentioned, viz. the non-linear character of its characteristic is sufficient to account for its unilateral conductivity, for, as H. Brandes points out, all conductors or combinations of conductors which do not follow Ohm's law, are capable of acting as detectors of electric oscillations owing to their rectifying effect. Hence, in any case in which the characteristic curve of a substance or its volt-ampere curve is not a straight line rising up from the origin, the material will be capable of acting as a rectifier for electric oscillations. Braun found that a number of substances, such as copper pyrites, iron pyrites, galena, and copper or antimony sulphide, possess unilateral conductivity, the current in some cases being twice as great in one direction as the other. He also found that no thermoelectric action could explain the phenomenon.

Nothing, however, seems to approach carborundum in this peculiar property. Pierce found that in one specimen platinised on one side so as to make an improved contact, the current under an impressed electromotive force of 34·5 volts was 527 times as great as the current in the opposite direction with the same voltage. In another case with 30 volts pressure the current in one case was 3000 or 4000 times greater in one direction than the other. As the current increases the efficiency of rectification decreases, but up to the present no theory has been proposed which satisfactorily explains this remarkable effect.

As regards the rectifying contact detectors, a very large number of these have been discovered, and they may be said to be the most widely used and popular form of detector known.

There are a number of pairs of substances, one generally a metal or good conductor, and the other generally a non-metal or often a sulphide or oxide of a metal, which when put in light contact offer greater resistance to the passage of an electric current one way than the other. Thus Dr. Austin found that a steel point in contact with a piece of silicon has this property. If an E.M.F. of 2 volts (D.C.) is applied, the current in a certain case was 3·2 milleamperes from silicon to steel, but 16 milleamperes from steel to silicon. In the same manner a carbon-steel junction has an unequal conductivity in the two directions. Again, a tellurium-aluminium junction is a good rectifier and also a plumbago-galena, the latter being a sulphide of lead.

A junction of a copper point with a mass of molybdenite (sulphide of molybdenum) rectifies well. A contact between gold and iron pyrites also rectifies, but almost the best couple is a contact of zincite (native oxide of zinc) with chalcopyrites (copper pyrites), discovered by W. G. Pickard, and called by him a *Perikon* detector. This rectification cannot be due to a simple thermo-electromotive force, because the rectified current is in many cases in the opposite direction to the true thermoelectric current generated by heating the junction.

Goddard has suggested that it is due to a film of oxide or sulphide which is more penetrable by ions of one sign than the opposite.

W. H. Eccles has assigned a thermoelectric origin to it, based on the possibly large electromotive forces due to the "Thomson" effect and Peltier effect, which occur in some of these imperfect conductors. The Thomson effect is, however, very difficult to measure, and hence it has not yet been possible to test this theory experimentally.

When inserted in series, with a telephone or galvanometer, such rectifiers obstruct the flow of the electric oscillation more in one direction than the other, and hence rectify the oscillation or convert it into a more or less continuous current or into gushes of electricity in one direction.

They are used in series with a telephone, and this circuit connected as a shunt across the condenser in the receiving circuit of a radiotelegraphic receiver taking the place of the Author's glow-lamp detector or rectifier as shown in Fig. 19. They are, however, more liable to get out of adjustment than the said glow-lamp detector. The Author has discovered that in all cases such rectifying contacts consist of two substances, one of which is more photoelectric than the other, that is, gives up electrons under the action of light more easily. Thus chalcopyrite is more photo-electric than zincite, and galena than plumbago. He has also found that when the substances are powdered up and then compressed again under great pressure into blocks or rods, the rectifying power has disappeared. It would seem, therefore, as if this rectifying power is due to a certain crystalline structure which is destroyed on crushing. This view is confirmed by the well-known fact that all points on the surface of such materials are not equally good in rectification.

It depends also in some degree upon the contact surfaces being unsymmetrical. G. W. Pierce has found that in the case of a slice of a carborundum crystal platinised on both surfaces, the rectification almost vanished. Hence the rectification in some degree depends on unequal contact surfaces. The full explanation of it has not yet been given, and all we can say is that it depends in some way upon the power of electrons at the contact surface to move more easily in one direction than the other when the contact is unsymmetrical and the material crystalline in structure.

It has been found possible to arrange a number of these rectifying substances in a series such that when taken pair and pair the largest current under a given external electromotive force flows from the substance lowest in the list to that which is highest in the list.

Thus, Mr. A. F. Hallimond has given the following list for 10 such substances which taken pair and pair give 45 different contacts. His list is in the order : Zincite, Brookite, Molybdenite, Zinc, Galena, Iserite, Copper, Chalcocite, Chalcopyrite, Tellurium. Thus, if we take zincite and chalcopyrite and connect the former to the negative pole of a cell and the latter to the positive and bring the pieces of mineral in contact, a larger current will flow than if the minerals change places.

The same holds good for any other pairs.

It appears, however, that zincite holds an important position and must possess some unique properties because it forms one element in all the really good rectifying couples. Nevertheless if this native mineral oxide of zinc is crushed to powder and then compressed again into a solid mass in an hydraulic press, the resulting slab of material possesses no rectifying power. Hence the property of permitting the flow of current better in one direction than the other cannot be due to any intrinsic quality in the chemical molecule, but must be the result of some special physical condition or crystalline structure.

8. **Electrodynamic Oscillation Detectors.** — Since high frequency alternating currents or electric oscillations create magnetic fields varying in a similar manner around the conductors through which they pass, and can induce oscillations of a similar frequency in neighbouring conductors, there are therefore attractions and repulsions produced between circuits through which oscillations are passing and others in which oscillations are induced by them, in virtue of the electrodynamic force between conductors conveying currents. We may, therefore, construct oscillation detectors which depend for their operation upon these forces of attraction and repulsion.

It was discovered independently by the author and by Elihu Thomson that if a metallic disc or ring is suspended by a fine wire within a circular coil through which an alternating current is passing, the ring or disc being held at an angle of 45 degrees to the plane of the winding of the coil, then the alternating current in the coil induces secondary currents in the disc or ring, and these create a mechanical force tending to make the disc turn, so that its plane is at right angles to the plane of the winding of the coil or parallel to the axis of the coil. It can be shown from first principles that when a closed inductive circuit is placed in an alternating magnetic field it will be acted upon by a torque, compelling it to move into a position in which it includes the least number of flux lines.

The author devised, as far back as 1887, an alternating current galvanometer depending on this principle, in which a copper or silver disc was suspended by a long fine wire in the above manner in the interior of a coil which could be traversed by alternating currents (see Fig. 22).

The same principle has been applied as a detector of electric oscillations, in 1899, by Fessenden, who used a suspended silver ring and two fixed coils on either side of it

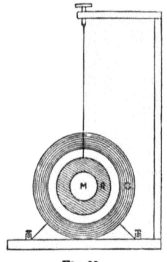

through which the oscillations pass. More recently, G. W. Pierce has increased the delicacy of the instrument by employing a disc of silver paper suspended by a long quartz fibre, the plane of the disc being hung at an angle of 45° to the axis of an ebonite tube, on the outside of which was wound a coil of insulated wire conveying the electric oscillations. A small fragment of silvered glass attached to the disc serves to reflect a ray of light upon the scale and to indicate a movement of the disc.

The average torque on the ring and therefore its deflecting moment is proportional to the square of the current in the coil and to the square of the frequency for the same instrument.

FIG. 22.

Hence, if the frequency is constant and the ring is suspended by a quartz fibre of constant size and length, the restoring torque varies as to the deflection, and the deflection would measure the mean square or integral value of the oscillations passing through the coils.

This conclusion has been confirmed by Pierce experimentally. This form of detector, therefore, like the thermal detectors, measures the integral value of the oscillations; but since the mechanical forces are small, such an electrodynamic detector is not nearly as sensitive as the best forms of thermal detector. Nevertheless, in some quantitative researches it has proved itself to be very useful.

9. **Mode of employing Oscillation Detectors in combination with Recording Instruments to detect Electric Waves.**—In considering the above-described oscillation detectors it will be seen that, with the exception of the electrodynamic detectors, they may all be divided into three classes—

(1) Those which under the action of electric oscillations undergo a change which in effect is equivalent to an alteration of resistance.

(2) Those which under the action of electric oscillations undergo a change which induces an electromotive force in another associated circuit.

(3) Those which possess a unilateral conductivity, offering therefore a greater resistance to the passage of a current in one direction than in the opposite direction.

As regards the first class, viz. those which under the action of electric oscillations undergo a change equivalent to a change in resistance, if we make the oscillation detector form a part not only of the oscillation circuit but of another circuit in which there is a continuous impressed electromotive force, and also some instrument capable of being affected by a change in this direct current, we can make the oscillation detector act the part of a relay in the following manner :—

The influence of the oscillations is to produce a change in the oscillation detector which virtually alters its resistance, making it greater or less. This action then in turn creates a change in the continuous current passing through it, generated by the continuous current appliance such as the voltaic cell, and this change in the continuous current is then able to affect some instrument capable of recording it by a visible or audible indication. Thus, for instance, an imperfect contact detector, such as the metallic filings coherer, undergoes a change under the influence of electric oscillations which cause it to become a better conductor. If, therefore, the metallic filings tube forms part not only of the oscillation circuit, but of another circuit containing a small unidirectional electromotive force, such as that provided by a voltaic cell, and some instrument capable of detecting a change in this current, such as a relay or galvanometer, then the influence of the oscillations on the metallic filings tube will cause it to undergo a sudden decrease in resistance, and therefore there will be an increase in the continuous current passing through it. This increase in the continuous current may be made to record itself permanently by employing some form of telegraphic relay.

A relay is a device by means of which a small increase in a very feeble current, or the passage through it of a very feeble current, is made to close another circuit containing a larger electromotive force and capable of carrying a larger current. A common form of relay consists of an electromagnet with coils of many turns. When a small current, say, of one milliampere, is passed through these coils the electromagnet attracts an armature, and this is made to close another circuit which then permits the passage of a much larger current, capable of working any form of telegraphic recording instrument. Thus one may associate together the following devices: A metallic filings coherer and a single voltaic cell may be joined in series with the electromagnet of a relay made as above described, and the second circuit of the relay may contain a battery of half a dozen cells and a Morse telegraphic printing instrument. If then the ends of the metallic filings coherer are connected with an antenna and balancing capacity, or with a pair of antennæ, and if electric waves fall in the right direction on this antenna they create electric oscillations which pass through the metallic filings and suddenly lower the resistance of the mass. At this moment the single cell passes an increased direct current through the coherer and through the magnet of the relay, causing it to attract its armature and in turn to close the circuit of the larger battery and printing instrument, and in this manner to record a mark on a paper tape. In place of the coherer, relay and a printing instrument we may employ some form of self-restoring detector in series with a telephone, and then the increase in the current through the oscillation detector due to its change in resistance under the action of oscillations suddenly alters the current flowing through the telephone, and it emits a sound which is heard as a short tick. These two methods are called respectively telegraphic and telephonic methods of receiving. Thus if in place of a metallic filings coherer we employ a thermal detector, then the action of the oscillations is to heat the fine wire of the detector and to increase its resistance, and if the wire is joined up in series with a single cell or telephone as above described, so that it is traversed not only by the oscillations but also by a continuous current due to the single cell, then the increase in temperature due to the oscillations decreases this continuous current suddenly, and also causes a sound in the telephone.

On the other hand, the magnetic detectors act in virtue of the change in magnetisation produced by the action of oscillatory magnetising forces brought to bear upon iron or steel. This change in magnetisation is made to create an induced electromotive force in another circuit embracing the iron or steel, and if this last circuit contains a telephone, the change in current through the telephone will give rise to a sound. Broadly speaking, we may say that the imperfect contact detectors and the thermal detectors and the electrolytic detectors experience under the action of oscillations a change which is equivalent to a change in resistance, whilst the magnetic detectors and the thermoelectric detectors act in virtue of a change which is equivalent to the production of an electromotive force in another circuit connected with the detector. In all cases the influence of the oscillations on the detector is made to bring about the increase or decrease of another current in another circuit, either by a variation of resistance or the introduction of an electromotive force, and this last current is made to indicate itself either on a telegraphic instrument, by the interposition usually of a relay, or else directly and audibly by means of a telephone.

In the case of rectifying detectors the process is somewhat different. Here the oscillation detector is of a material, either rarified gas or crystal, which has a unilateral conductivity, and converts the oscillations directly into a continuous current, capable of being appreciated and indicated either by a galvanometer or by a telephone.

Taking the whole arrangement together, the oscillation detector and the associated circuit and the antenna connected to the oscillation detector, we have a sensitive appliance for detecting the passage of electric waves through space, which may therefore be called a *cymoscope* or wave detector when so used.

In reviewing the action of oscillation detectors generally, we remark that in some of them the energy of the oscillations created in the receiving circuit is allowed to expend itself directly in affecting some indicating instrument, such as a telephone or galvanometer. In other cases the energy of the oscillations merely releases the energy of some external source, and it is this which affects the indicating instrument. In these last cases the action is called a trigger action, because it is similar to the operation by which the pressure of the finger on the trigger releases the energy of the powder which in turn propels the bullet. The coherer, electrolytic detector, and magnetic detector are instances of this last class, whereas the thermal detector, gaseous and crystal rectifiers are instances of the first class.

We may in conclusion make a brief reference to methods for the quantitative measurement of electric waves. An important appliance in this connection is the form of galvanometer called an Einthoven galvanometer. This consists of an extremely fine quartz fibre which is silvered by depositing on it chemically an extremely thin layer of silver. This fibre is stretched in the very powerful magnetic field of an electromagnet transversely to the direction of the lines of force. The pole pieces of the magnet have holes bored through them so that the fibre can be viewed by a microscope, or its image projected optically upon a screen or else upon a strip of photographic paper. In series with the fibre is placed a crystal of carborundum or else a Fleming oscillation valve or some other rectifying contact, so as to convert the electric oscillations into a unidirectional current. When this rectified current flows through the fibre it causes it to be deflected or displaced in the magnetic field, and this displacement is proportional to the strength of the current, other things being equal.

The instrument is capable of measuring a current of a few microamperes or less. Hence it is very suitable for the measurement of such oscillations as are produced by electric waves falling on an antenna.

For this purpose it is necessary to calibrate the Einthoven galvanometer, using damped electric oscillations of the same spark frequency and oscillation frequency as is employed in the wave generation for the waves under test.

A convenient method of doing this is as follows: Two flat spiral coils are arranged so that they can be moved to or from each other with planes parallel.

The first step is to find the mutual inductance of these coils at various distances apart. Having done this we know the relative strengths at various distances of the secondary currents induced in one spiral by damped oscillations in the other.

Hence if we create in one spiral electric oscillations of any required period and group frequency, we can, by putting the spirals very close to each other, obtain a secondary current of such strength that it is possible to measure its R.M.S. value on a hot wire ammeter which has been previously calibrated with continuous currents.

If, then, we move the spirals to any greater distance apart such that the induced rectified oscillations can suitably deflect an Einthoven galvanometer, we can calculate the R.M.S. of this current, and therefore calibrate the Einthoven.

In this manner it is possible to determine the variation in strength of wireless signals received at various times from any transmitting station at a receiving station, provided we can imitate by some oscillation producer the type of oscillation used in creating the waves as regards oscillation and group frequency.

In the next chapter we shall consider the use of these appliances in wireless telegraphy, and also mention some other devices especially adapted for the detection of undamped electric waves.

CHAPTER VII

RADIOTELEGRAPHIC STATIONS

1. **The General Principles of Radiotelegraphy.**—Having described in the previous chapters the methods employed for generating electromagnetic waves and radiating as well as detecting them at distant places, we have next to consider the combination of these processes into a practically operative system of radiotelegraphy.

To convey information to a distance we must be able to produce at the receiving station at pleasure certain visible or audible signals signifying letters, words, or ideas. For this purpose the most commonly used code is the International system of Morse signals, according to which each letter of the alphabet is denoted by a collocation of elementary signals of two kinds, one of short length or duration, called a *dot*, and the other of three times the length or duration, called a *dash*. Groups of these dots and dashes are made to succeed each other, with an interval equal to the length or duration of a dot between them, to form the various letters or numerals. Thus, the International Morse Code usually employed is as follows :—

THE ALPHABET.

THE NUMERALS.

1 ▪ ▬ ▬ ▬ ▬	6 ▬ ▪ ▪ ▪ ▪
2 ▪ ▪ ▬ ▬ ▬	7 ▬ ▬ ▪ ▪ ▪
3 ▪ ▪ ▪ ▬ ▬	8 ▬ ▬ ▬ ▪ ▪
4 ▪ ▪ ▪ ▪ ▬	9 ▬ ▬ ▬ ▬ ▪
5 ▪ ▪ ▪ ▪ ▪	10 ▬ ▬ ▬ ▬ ▬

Full stop ▪ ▪ ▪ ▪ ▪ ▪   Understand ▪ ▪ ▪ ▬ ▪

Repeat ▪ ▪ ▬ ▬ ▪ ▪   Call Signal ▪ ▪ ▪ ▬

If, then, we have the means of marking upon paper a collection of these signs or making them audible as short and long sounds in a telephone, we can signal out letters, and, therefore, words. A space equal to a dash is left between letters and a longer space between words.

In the case of non-alphabetic languages, like Chinese and Japanese, the ideographs are numbered, and the numbers transmitted and translated. If the Morse characters can be printed on paper strip as received, we have a permanent record. If they are received by telephonic sounds, the observer translates them mentally, and writes down the letter on paper as received. Accordingly, the broad principles of all radiotelegraphy are as follows:—

At one place, called the *transmitting station*, there must be an antenna or radiative circuit, called the sending antenna or radiator, and means must be provided for creating electric oscillations in this circuit, which may be damped or undamped. A circuit closer, called a sending key, must be included in the transmitting circuit, by which the oscillations or successive trains of oscillations can be started or stopped at pleasure, and hence electromagnetic waves radiated in trains or groups of trains, of shorter or longer duration, to correspond to the signals of the Morse alphabet or any other similar signal code.

At some other place, called the *receiving station*, there must be a similar antenna or absorbing circuit to absorb these waves. In the course of this circuit, or connected with it, there must be some form of oscillation detector to be influenced by the oscillations set up in the absorbing circuit by the impact on it of the waves sent out from the corresponding transmitter. This oscillation detector must have some appliance, such as a telephone or telegraphic recording instrument, connected with it, which it influences, and then there must be an observer to hear or note the signals so given, which correspond with those made by the sending key at the distant place.

By this means we transmit signals without continuous wires, by means of electromagnetic waves, from one place to another, which are interpretable as alphabetic or intelligible signals conveying information, and this constitutes the art and practice of radiotelegraphy.

The precise appliances employed differ according to the distance of the stations and their locality. The particular characteristic of radiotelegraphy, as compared with conductive or fixed wire telegraphy, is that one or both of the stations may be in motion. Hence, it forms an ideal means of communication between two ships at sea or between ships and the shore. Also, since sea water is a good conductor relatively to dry land, communication by radiotelegraphy over sea is facilitated, as already explained, and it is therefore especially marked out as a means of supermarine communication. Nothing is more remarkable than the rapid rate at which the maximum distance of communication over-sea by radiotelegraphy has been extended. Prior to 1896 no one had been able to demonstrate the detection of an electromagnetic wave at a greater distance from its generator than about half a mile. In 1897, Marconi gave the first demonstration in England of actual radiotelegraphy over a distance of several miles, and in 1898 had an operative system at work between Bournemouth and the Isle of Wight, a distance of about twelve miles. In 1899 he accomplished the feat of radiotelegraphy across the English Channel, and for the first time drew public attention strongly to the possibilities of the new telegraphy. By the end of that year, or the beginning of 1900, a distance of 100 miles was covered by him, and inventors all over the world were endeavouring to follow him in these achievements.

At the beginning of 1901 he had signalled in this manner 200 miles, from the Isle of Wight to Cornwall, and at the end of 1901 had succeeded in sending alphabetic signals 3000 miles over the Atlantic Ocean. Since that date its achievements have steadily been extended by him and others, and in 1908, or seven years from the transmission of the first transatlantic signals, regular radiotelegraphic communication was established between Ireland and Nova Scotia by the ingenuity and perseverance of Mr. Marconi and those associated with him. By that date also every navy in the world had adopted it as an indispensable means of signalling.

We shall, then, in the following sections consider in detail the apparatus now employed for short and long distance radiotelegraphy, taking the various elements of the apparatus in succession, and dealing first with that employed for short distances, viz. up to 100 or 200 miles, both on shore and on ship, for the purposes of supermarine intercommunication.

2. **Short Distance Radiotelegraphic Apparatus. Antennæ or Radiator Supports.** — In establishing a short distance radiotelegraphic station, the first question to be considered is the site and the erection of a support for the antenna. Another important matter is the possibility of securing a "good earth,"

FIG. 1.—Marconi's Wireless Telegraph Station at Poole, Dorset.

in a telegraphic sense. As the greater part of short distance radiotelegraphy is conducted over sea and is concerned with communication with ships, such a station is nearly always established on the coast. It is desirable that the soil at the point selected should not be too dry or rocky, as hard dry rocks are poor conductors, and render it difficult to obtain the necessary earth connection. It is also desirable that the site should not be overtopped by hills, and that it should have an open outlook to the sea in the direction in which the transmission is chiefly desired. The soil, however, should be sufficiently firm to enable good foundations to be obtained for the mast or masts used as the antenna support. Assuming the station to be for short distance work, a single mast generally suffices for this purpose. This is usually erected in three sections, 50- or 60-feet poles being employed, each of which must be well stayed, and if galvanised iron wire is employed for the stays, these should be interrupted by insulators at intervals, so as to avoid having long wires which might have a natural time-period of oscillation equal to that of the antenna employed. At the top of the mast a gaff is erected with pulley and tackle for hauling up the antenna wire (see Fig. 1).

In the case of ship installations it is usual to provide either the main or fore mast with an additional gaff to carry the antenna. The antenna itself is preferably constructed of hard drawn tinned copper, phosphor bronze, or aluminium wire, and it is nearly always in the form of a multiple wire antenna. If a fan-shaped multiple wire antenna is employed, then it is generally necessary to erect two masts with a horizontal or triatic stay between them, from which the antenna is suspended (see Fig. 2), but if a single mast is employed, then the antenna may be of a double cone, or preferably of the umbrella form (see Fig. 11, Chap. V.). When the spark system is employed, then, owing to the high potential of the extremities of an open antenna, the upper end must be extremely well insulated. A form of insulator devised by the author, suitable to this purpose, is made as follows: A thick-walled ebonite tube, about 2 feet or 60 cms. in length, has a brass wire down the centre ending in a loop at the bottom and a button at the top. The top button is covered over with an ebonite cap fitting perfectly watertight. The ebonite tube is gripped by a cross oak bar just below the top, and below this a conical or tubular rain shield is attached watertight to the top, so as to

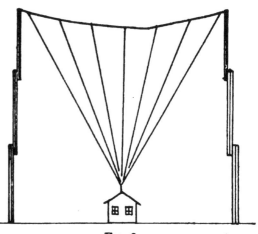

Fig. 2.

keep the ebonite as far as possible dry. One or two such insulators may be used in series. Similar insulators are used to draw out the wires into desired positions.

A closed or loop antenna may be constructed of two or more antenna wires connected together at the top and sustained by an insulator, if required, the middle or some lower points of the two wires being pulled out by rope stays attached to insulators, so as to form a lozenge-shaped or triangular closed or loop antenna.

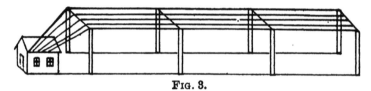

FIG. 3.

In cases in which bent or directive antennæ of Marconi's form are employed, then two or more masts have to be erected to sustain the horizontal portion of the antenna in the required direction (see Fig. 3). A very common form of antenna is that called an "umbrella antenna." In this case a single mast or tower is erected, and from the top insulated wires extend outwards

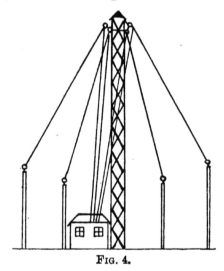

FIG. 4.

and downwards like ribs of an umbrella (Fig. 4). The tower is sometimes insulated at the base and forms part of the antenna. At other times the wires are insulated at the top from the tower.

Unless the mast or masts are erected in contiguity to some already existing building or lighthouse, it is then necessary to erect near the base of the mast or masts in some convenient position a building which serves as a radiotelegraphic station, which may be of wood, brick, or galvanised iron, as most convenient. Into this building the antenna is then led, and an important detail of construction is the means necessary to secure the necessary insulation and yet weather tightness. One method is to cut a hole in a pane of glass in a window, or to make a window glazed with a single sheet of thick plate glass. In this a hole is bored, through which passes a thick-walled ebonite tube made tight with indiarubber washers. It is advantageous to protect this glass from the weather as far as possible by an overhanging hood facing in the direction opposite to that of the prevailing wind. Through the ebonite tube passes a thick stranded copper wire cable well insulated with indiarubber. The outer end is connected to the bottom of the antenna, and the inner end is brought to a highly insulated switch, by means of which it can be connected as required to the transmitting or receiving apparatus. In most cases the same antenna serves alternately as a sending and as a receiving antenna. In some cases simultaneous transmission and reception is provided for by the use of two antennæ. The receiving antenna is then given such a form as to be uninfluenced by radiation from the adjacent sending antenna.

As already observed, an important element is the earth-plate or connection to earth, which is unquestionably necessary for effective radiotelegraphic communication for any considerable distance. This is made by burying in the earth strips of sheet copper or sheet zinc, or thick copper wires radiating from a convenient point like the roots of a tree. Galvanised iron should not be used, as it soon corrodes in damp soil. The object should be to obtain the greatest possible surface exposed to moist earth, as in the case of a lightning conductor, and in order to be able to test the resistance of this earth, it is desirable to construct this earth-plate in three parts separated from one another. The resistance between the three portions can then be measured by a Wheatstone's bridge or any other method, but if a single earth is employed the earth resistance cannot easily be measured. In any case a thick stranded cable or wide copper strip must be connected to the earth-plate or earth-wires and brought into the sending and receiving room of the radiotelegraphic station; and during all times when the apparatus is not being used, the antenna should be connected to the earth-plate directly, so as to form a well-earthed lightning conductor.

A thoroughly good antenna should resemble a tree, in having

almost as much beneath the ground as above it, the trunk and branches corresponding to the exposed portion of the antenna and the ramified rootlets to the earth-wires.

In the case of a ship no difficulty exists in obtaining a good earth, because a connection can be made to the copper sheathing or to the outer plates of the ship.

In dry places it is desirable to make arrangements for putting water down on the earth-plate, so as to keep the soil round it moist. One of the great difficulties of effecting radiotelegraphy in inland places in many tropical countries is the dryness of the soil and the difficulty of getting a thoroughly good earth connection.

It is possible, however, to operate without an earth connection. It is then usual to make a balancing capacity be spreading out above the earth a number of radiating wires which are near to, but insulated from the earth. This forms with the earth a condenser, or so-called balancing capacity.

3. The Arrangement of the Transmitting Apparatus for Short Distance Radiotelegraphy.—In the case of radiotelegraphy conducted with damped electrical oscillations, or so-called spark telegraphy, the transmitting apparatus comprises three elements.

(1) Some means for charging the antenna or other condenser to a high potential.

(2) A discharger permitting this charge to flow out of the condenser with oscillations which are either communicated directly or inductively to the antenna.

(3) Means for controlling these oscillations or repeating them in long or short groups in accordance with the Morse signals.

For short distance spark telegraphy the necessary high potential is always obtained by the use of an induction coil or transformer; either a single instrument or a number of induction coils or transformers may be employed, having their secondary circuits joined in series and their primary circuits in parallel.

The most usual appliance is an induction coil of the ordinary type giving a spark 30 to 60 cms. and taking current either from primary or secondary batteries or from a small alternator. The induction coil is placed on a table in the transmitting station, or it may be fastened to the wall. It may be operated either by alternating currents or by an interrupted continuous current. If operated by a continuous current, this may be taken from a battery or a dynamo.

In isolated places and on lightships and in lighthouses it is usual to employ secondary batteries which are charged by primary batteries. The ordinary 10-inch induction coil usually requires a primary current of 10 amperes at 16 volts. Hence, eight or ten secondary cells are sufficient to work it. These, however, must be charged by a battery of primary cells, say, 20 primary cells in series, each giving 1·5 volts E.M.F. One arrangement which may be employed is to connect up 100 large dry cells, five in parallel and twenty in series, and join them in parallel with ten secondary cells, and connect the ends of the secondary cells to the terminals of the primary circuit of the induction coil. The primary cells then charge the secondary cells, and the secondary cells give up current to the coil as required.

In land stations a convenient arrangement is a small oil engine and continuous current dynamo, which is employed to charge the secondary cells, which in turn are used with the coil. On board ship the current can be taken from the lighting circuits, which generally furnishes a continuous current at 50 or 100 volts. In cases where the continuous current is employed, some form of interrupter must be used with the coil. The ordinary hammer break with platinum contacts is still used on board ship on account of its simplicity, but mercury turbine breaks are also much employed. In this latter case it is usual to suspend the mercury break in gimbals. The Wehnelt or electrolytic break is not often used, as it involves the use of acid and is troublesome to keep in order and somewhat irregular in action. The bulk of the work now done may be said to be divided between the hammer break with platinum contacts and the coal gas mercury break.

. The next element in the transmitting apparatus is the signalling key, for interrupting the primary circuit in accordance with the signals of the Morse alphabet. This must be a quick break key with a long ebonite handle easily operated, and has generally a magnetic blow-out in connection with the platinum terminals between which the interruption takes place. When alternating currents are employed, a key is employed by means of which the circuit is only actually interrupted at a time when the alternating current passes through its zero value. This is achieved by means of an armature, which when once pressed down by the movement of the hand, keeps the circuit closed until the alternating current in the course of its cycle passes through its zero value, when the key automatically opens its circuit again. In cases where alternating current is supplied, or can be provided by means of a special alternator, it is usual to substitute a closed iron circuit transformer, insulated in oil, for the induction coil, and this is also done on board ship in the case of high power transmitters for long distance work.

We have next to consider the methods of creating the oscillations in the antenna. The simplest way of doing this is by employing the induction coil to charge the antenna directly, as in Marconi's original invention. (See Fig. 5). In this case, the antenna, A, is connected to one of the secondary terminals of the induction coil, I, the other terminal being connected to the earth, E, and the two terminals also joined to a pair of spark balls, S, which can be more or less approximated. At each interruption of the primary circuit an electromotive force is created in the secondary circuit, which charges the antenna, and when this potential difference reaches the spark potential corresponding to the distance at which the spark balls are placed, the antenna discharges itself across the gap with highly damped oscillations. This direct charging of the antenna is now never employed in practical work except perhaps for experiments.

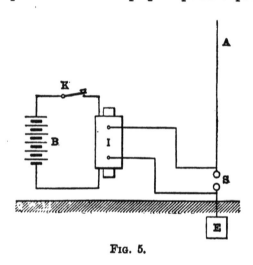

FIG. 5.

The most usual way of charging the antenna is by connecting it directly or inductively to a closed condenser circuit interrupted only by the spark gap. In this case the secondary terminals of the induction coil, I, or transformer are connected to the adjustable spark balls, S, and these balls are also connected by a condenser, C, consisting usually of a battery of Leyden jars, and by an adjustable inductance coil, L. The antenna, A, is connected directly to one point on this circuit (see Fig. 6), some other point being connected to earth. The antenna and condenser circuits must be syntonised. This may be done by connecting in series with the antenna another adjustable or tuning inductance by means of which the natural time period of the antenna is made to agree with that of the condenser circuit. This tuning may be achieved by connecting a hot wire voltmeter over one or two turns of the inductance in series with the antenna, and then altering this inductance, or else altering the capacity and inductance in the condenser circuit until the indications of this voltmeter are a maximum. Marconi prefers to connect the antenna, A, inductively with the energy-storing circuit by means of an oscillation transformer called a transmitting jigger. This jigger consists of a wooden frame on which are wound two circuits, *p, s,* a single

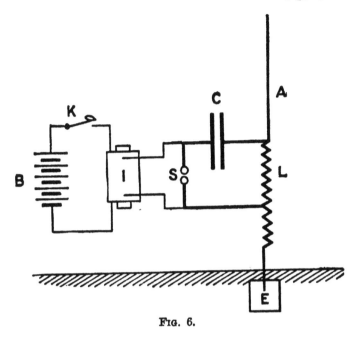

turn or two of a small inductance which forms part of the condenser circuit, and another overwound circuit 5 to 10 turns in

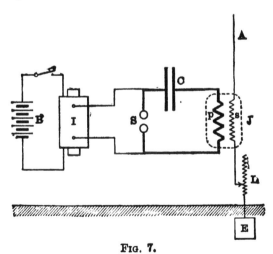

series with the antenna, and with a tuning inductance, L (see Fig. 7).

This last method possesses the great advantage that the closeness of coupling of the antenna and condenser circuit can be varied.

It has already been shown in Chapter I., that when two oscillation circuits are connected together inductively, and in tune with one another, oscillations set up in one circuit result in the production of oscillations in both circuits, having two frequencies, one greater and the other less than the natural frequency of each circuit when separate. Hence, when we are employing an inductively coupled antenna which has been syntonised with the condenser circuit, and an unquenched spark discharger, oscillations of two frequencies are set up in the antenna, and waves of two wave lengths radiated from it, one greater and the other less than the wave length corresponding to the natural frequency of the antenna taken alone. One of these waves has greater amplitude than the other, the longest wave length being the least damped, and, therefore, generally speaking, having the greatest integral value. The wave lengths approximate to one another in proportion as the coupling is made weaker, but then they also diminish in amplitude, so that by the employment of a weak coupling, which can be done by separating the primary and secondary circuits of the oscillation transformer, we gain a radiation which is relatively feeble, but of one single wave length, whereas by coupling closely together the primary and secondary circuit of the oscillation transformer we have more powerful waves but waves of two wave lengths radiated, and the receiving antenna must accordingly be syntonised to one or other of these wave lengths.

As regards condensers for the oscillation circuit, although a Leyden jar is a bulky form of condenser in comparison with its energy-storing power, nevertheless its simplicity still recommends it. The main condenser often consists, then, of a battery of Leyden jars, a certain number being joined in parallel or partly in parallel and partly in series. It is very important that these jars should have their capacity marked upon them, and that they should be selected so as to be exactly equal. When oscillations are taking place in the condenser circuit, it will be seen that an electric brush discharge takes place from the free edge of the outer tinfoil. It has been shown by Eickhoff that this brush discharge involves a considerable expenditure of energy, and, therefore, tends to damp out the oscillations. The author has also shown that it increases the capacity of the jar by an irregular amount. Both these effects can be stopped by putting the jars in insulating oil, and in place of Leyden jars it is in every way preferable to employ a condenser consisting of sheets of glass or ebonite coated with metal and immersed in oil, or the glass and ebonite may be omitted, and the condenser constructed simply of metal plates immersed in oil. Brush discharges are thereby prevented, and accuracy of tuning is secured by preserving a constant known capacity in the condenser circuit. In the case of either the inductive coupling or the direct coupling of the antenna to a condenser circuit, it is of the greatest importance to secure an exact syntonisation between the antenna and condenser circuit, as a very little want of syntony immensely decreases the strength of the current flowing into and out of the antenna. Other things being equal, the radiation from the antenna will be proportional to the mean square value of the current flowing into the base of the antenna. This current may be measured by inserting in that point a hot wire ammeter.

Another important element in the transmitting arrangement is the spark discharger. Where large capacities are being employed, the noise of this spark is very distressing, and moreover it enables the messages to be. read at a great distance by any one familiar with the Morse alphabet. Hence, for many reasons, the spark balls should be enclosed, as first suggested by the author, in a cast-iron chamber with a glass peephole in it. The chamber is preferably kept full of nitrogen or carbonic dioxide gas, and should also have some lime or alkaline material in it for absorbing acid vapours. In this manner the discharger can be rendered perfectly noiseless. It is a great advantage to blow a jet of air upon this spark gap to quench the arc. This can easily be done by means of a small Lennox blower operated by an electric motor.

In the case of short distance transmitting plant, the discharger generally consists of a pair of balls of iron or brass, with arrangements for turning them round in such a way that every now and again fresh surfaces can be brought into apposition, as the continual action of the spark erodes or cuts away material from the spark balls, leaving them rough. The general appearance of a Marconi ship transmitter is as shown in Fig. 8, which gives a view of the Marconi transmitting apparatus used on board ship, the set comprising a motor-alternator for producing the alternating current which is seen below the table, and the spark balls, condenser and oscillation transformer on it.

At one time it was considered an advantage to divide the spark up between several spark surfaces by putting balls in series, but this does not appear to be the case, and it is usual now to employ a single spark gap. A view of a fixed single spark gap is shown in Fig. 9.

When an inductive coupled transmitter is being used consisting of a radiator or antenna coupled to a condenser energy storing circuit, as shown in Fig. 7, then as long as the spark at the spark balls endures the primary condenser and inductance forms a closed

Fig. 8.—View of Marconi Ship Wireless Telegraph
Transmitting Apparatus.

circuit, and the reaction between the current in it and the current induced in the open antenna circuit gives rise to the production of a complex oscillation as already explained, which is resolvable into oscillations of two frequencies. To avoid this division of the energy it is now usual to employ some form of quenched spark gap which has also the property of producing spark discharges of high frequency and of extremely regular occurrence to form what is called a musical spark.

It has been already mentioned that when two smooth thick copper plates are placed very near to each other, say half a millimetre, with surfaces perfectly parallel, a condenser discharge made to take place between them, the discharge does not endure, but is soon quenched or arrested by the cooling action of the masses of metal forming the discharge surfaces. Hence, if a number of such plates are placed parallel to each other with their surfaces truly plane and parallel, a condenser discharge taking place through the series of plates will have to jump over a number of narrow air-gaps and will be quickly quenched or arrested. If, then, such a multiple plate quenched spark discharger is placed in the primary condenser circuit, the condenser discharges which take place across it are nearly dead beat or contain but few oscillations in a train.

The primary circuit, therefore, soon becomes open, and the

Fig. 9.—Enclosed Spark Gap in Silencing Chamber (Marconi).

effect on the secondary or antenna circuit is to give it a sudden electrical impulse, and to set up in it prolonged trains of oscillations of its own natural frequency.

The flat plate discharger has, however, the disadvantage that there is nothing to determine or produce any regularity in the intervals of time between two discharges. Hence invention has been directed to the production of rotating spark dischargers which can produce a rapid and very uniform series of quenched sparks.

Two such forms of musical quenched spark dischargers have been invented by Mr. Marconi, and in connection with all the ship or shore transmitters, involving anything but the smallest power, a rotating spark discharger of the following kind is now used by the Marconi Company.

The alternating current is supplied by an alternator giving a frequency of about 300 or so. This alternator is driven by a direct current motor which takes current from the ship's electric lighting circuits.

On the shaft of the alternator is placed a wheel without rim with insulated hub, the spokes being metal rods connected together metallically. This wheel has near its circumference two insulated metal terminals so that in its rotation the spokes just graze these terminals and connect them electrically (see Fig. 10). The wheel and terminals are enclosed in a box which is traversed by a current of air produced by a fan.

These insulated terminals form the extremities of a circuit containing the main condenser and the primary circuit of the oscillation transformer. Hence as the wheel rotates it discharges the condenser at regular intervals, and the spokes are so placed that the condenser is discharged just when the alternating electromotive force of the alternator reaches its maximum value during the period. The spark is then extinguished almost directly and gives a quenched spark discharge in the condenser circuit, and thus sets up in the sending antenna trains of free oscillations of one single period spaced at very regular intervals. This type of discharger is shown in another form in Fig. 21, and similar dischargers of a slightly different type are described later in this chapter.

So far we have only considered the use of damped oscillations for transmitting, with which at present (1915) the bulk of the short distance radiotelegraphic work of the world is conducted. If undamped waves are employed, then a Poulsen arc takes the place of the spark balls.

The copper-carbon arc apparatus, already described in Chapter III., would have its electrodes connected by a small capacity consisting of metal plates in oil and a large inductance coil, this oscillation circuit being so arranged as to have a frequency of something of the order of a million. The antenna is generally directly connected to this oscillation circuit and tuned to it. The signals are then made by short circuiting, by means of a key, a few turns of this inductance so as to throw the oscillation circuit out of tune with the antenna. The arc has to be fed from a continuous current dynamo giving an E.M.F. of 400 to 500 volts,

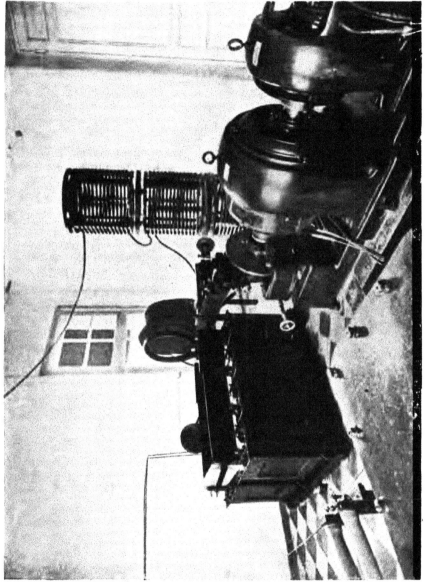

Fig. 10.—Marconi Transmitting Plant, showing the Motor-Alternator with High-speed Musical Discharger on end of Alternator Shaft.

suitable regulating resistances being interposed. The arc is enclosed in a metal box with cooling flanges attached, clockwork or an electric motor being employed to rotate the carbon. In place of hydrogen or coal gas the vapour of alcohol is now used to exclude the air. Alcohol is admitted drop by drop to the arc chamber from a sight feed lubricator attached to it (see Fig. 11).

For short distance work an arc formed with 220 volts and a small current of 2 or 3 amperes is sufficient, but for long distance work an arc taking 8 to 10 amperes at 440 volts is employed. In this last apparatus the magnetic field in which the arc burns is provided by a large electromagnet on which the arc box is fixed (see Fig. 11).

4. **Short Distance Receiving Apparatus.** —Turning next to the receiving arrangements, we find that they differ chiefly in the nature of the actual oscillation detector employed and the recording or signal making appliance.

[*By permission of the Amalgamated Radiotelegraphic Co., Ltd.*

FIG. 11.—Poulsen Arc Apparatus.

It is a very great advantage to possess a permanent automatic record of the signals made by some form of telegraphic instrument. On the other hand, the mechanical and electrical inertia involved in some appliances limits the speed. We may avoid this, and increase the speed by using a telephone with a self-recovering oscillation detector or a photographic recorder. Accordingly, the receiving apparatus may produce self-recording visible signals on telegraphic paper tape, or audible signals by means of a telephone. In the first case, some kind of imperfect contact oscillation detector may be employed in conjunction with a telegraphic relay, and printing telegraph or recorder. In the other case, a magnetic, electrolytic, or rectifying detector with a telephone is used, or a thermoelectric detector in combination with a photographic recorder.

The arrangements for telegraphic reception employed by Marconi in the early days of wireless telegraphy were as follows :—

The antenna, A, is connected to earth through the primary coil, j_1, of an oscillation transformer, and a tuning coil is inserted in between the earth plate and this primary coil. The secondary coil, j_2, of the oscillation transformer is cut in the middle and

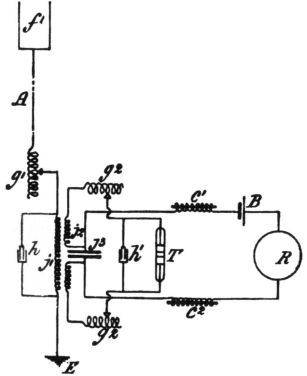

Fig. 12.—Arrangement of an early form of Marconi Wireless Telegraphic Receiving Apparatus, employing a Coherer, Relay, and Printer.

a condenser inserted. The outer ends of the secondary coil are connected to a Marconi metallic filings coherer tube, T. The terminals of the tube are also connected by a second or tuning condenser (see Fig. 12). The ends of the first-named condenser are connected through two small choking coils, C_1, C_2, with the electromagnet circuit of a telegraphic relay, R, and with a couple of dry cells, B, so that when the coherer tube is made conductive by the oscillations a current from these cells flows through the tube and actuates the relay. The relay was then made to work a Morse inker by means of 6 or 8 more cells. The whole apparatus, coherer tube, tapper, relay, oscillation transformer, condensers and cells, is mounted up on a stout baseboard, and enclosed in a metal box with sliding door so that the operator can get his hand in to the numerous set screws which adjust the relay and tapper to their best positions. The relay used is the so-called polarised relay. In this a horse-shoe electro-magnet is fixed on one end of a permanently magnetised steel bar, the other end of which is bent round and carries a delicately pivoted steel tongue, which is held with its free end between two soft iron pole pieces fixed on the ends of the electromagnet. The steel tongue then tends to stick to one or other of these poles. It is pulled away from one pole by an adjustable spring, and is held in balance against a stop, so that if a very small current is sent through the coils of the electromagnet, it magnetises one of the soft iron poles North and the other South, and causes the movable tongue to fly over against the other platinum point stop and make a contact with it, whilst at the same time it closes a second electric circuit.

In the case of relays used on board ship, the tongue must be so balanced that there is no tendency for it to move merely by the pitching or rolling of the ship, and the relay must furthermore be enclosed in an airtight box to prevent the damp sea air depositing moisture on the platinum contacts.

Such a relay will generally have a resistance of 1000 ohms or more in its electromagnet coils, and will work with a current of 0·1 of a milliampere and close a circuit, enabling a current of 0·1 ampere or more to be passed through the other circuit which is closed by the relay. The adjustment of the relay is effected by turning a screw which moves over the pair of soft iron poles together one way or the other, and so alters the pressure of the tongue against one of its stops.

The recording instrument sometimes used in telegraphic receiving is called a Morse inker, and consists of a clockwork mechanism which drives a strip of paper tape under a roller at a uniform speed (see Fig. 13). At one end of the instrument an electromagnet, M, acts upon an armature carried on the end of a lever, to the other end of which is attached an inking wheel, W, which dips into a well of ink, I, so that when the armature is attracted by the magnet the wheel is pressed up against the underneath side of the paper tape, p, and marks upon it a short or long line, according to the time during which the electromagnet is excited. Hence, when the relay closes the circuit of a local battery, B, which last is in series with the magnets of the Morse inker, a mark will be made upon the paper.

At the same time the relay closes the circuit of another electro-magnet, to the armature of which is fixed a hammer like that of a trembling electric bell. This tapping magnet is so arranged that the hammer strikes the underneath side of the coherer tube, and there are adjusting screws which regulate the range and frequency of the blows. The whole operation taking place in the receiver is then as follows :—When an electric wave falls upon the antenna it excites oscillations in the antenna. These are transformed by the oscillation transformer or receiving jigger, as it is called, and these oscillations pass through the coherer tube, causing it to become conductive. This then passes a current from the single or double cell through the relay magnet, and actuates the relay. This last in turn closes the circuit of the larger battery and sets in operation the Morse inker and the trembling tapper, which at

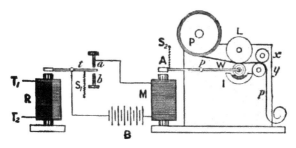

FIG. 13.—Scheme of Connections of Relay and Morse Inker.

R, Relay magnet.
*T*₁, *T*₂, Terminals of relay.
a, *b*, Relay contacts.
t, Relay tongue.
M, Morse inker magnet.

B, Local battery.
W, Ink wheel.
P, Paper strip wheel.
I, Ink well.
p, Paper tape.

once taps the coherer tube back to a condition of non-conductivity, and at the same time a mark is made upon the paper as long as the waves or trains of waves continue to fall upon the antenna. A number of adjustments are necessary to get a good result. The relay must be adjusted to proper sensibility, and also the magni-tude and force of the blow given by the trembling hammer on the underneath side of the coherer tube, by the adjusting screws. Also it is necessary to regulate the sensitiveness of the Morse inker by altering the tension of a spring which pulls the armature away from the electromagnet. The apparatus requires a certain skill in management, and the training required in an operator is not merely that of learning to send accurately and quickly the Morse signals with proper spacing by the transmitting key, but much more in adjusting the numerous regulating screws of the receiving apparatus, so as to make it record accurately and unintermittently the signals received.   In the Lodge-Muirhead receiver using the self-restoring mercury and steel disc detector described in Chapter VI. no tapper is necessary.   By means of a shunted voltaic cell an electromotive force of a fraction of a volt is applied between the mercury and the wheel through the circuit of a Kelvin syphon recorder.   When oscillations are passed across from the mercury to the disc they temporarily break down the insulation of the oil film and the external E.M.F. then causes the glass pen of the syphon recorder to make a deflection, which, however, subsides as soon as the oscillations cease.   Hence, on the paper tape the uniform straight line drawn by the pen is broken by a sudden sharp notch or hump representing a *dot* and a larger square-shouldered hump representing a *dash*, according as the pen is kept deflected for a shorter or longer time.

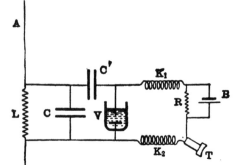

FIG. 14.

Although there is a great advantage in possessing a permanent record on the tape, the numerous adjustments then necessary and the reduction in speed of reception due to mechanical and electrical inertia in the receiving arrangement have caused this coherer receiver to fall out of use.   What is now generally employed is a telephonic method of reception, which possesses much greater simplicity.   In this case, the receiving apparatus consists of the antenna and the oscillation transformer, the primary circuit of which is inserted in the antenna circuit, and the secondary circuit connected to a condenser.   To the terminals of this condenser are connected the electrolytic receiver or rectifying detector employed, and to the terminals of the said detector are also connected a circuit including a telephone and a single voltaic cell, shunted by a variable shunt (see Fig. 14).   In this case, the oscillations set up in the antenna, A, give rise to secondary oscillations in the circuit comprising the tuning condenser, C, and the secondary circuit of the oscillation transformer, and these oscillations, when they have reached a certain amplitude, act upon the electrolytic receiver, V, and cause its resistance to fall.   The local voltaic cell, B, then sends a current through the electrolytic cell and through the telephone, T, causing a sound in the latter; but as soon as the oscillations cease in the antenna, the electrolytic cell rises again instantly to its former resistance. When a rectifying crystal or contact is employed the trains of oscillations are rectified into gushes of electricity all in the same direction, and these create in the telephone a musical sound the pitch of which depends on the spark frequency. Accordingly, if trains of oscillations fall upon the antenna, a rapid succession of short sounds is made in the telephone, which run together into a sound of continuous duration, prolonged as long as the trains of waves fall on the antenna. Hence, audible signals, corresponding with the dot and dash of the Morse alphabet, can be made by a suitable emission of long or short groups of trains from the transmitting antenna. The local receiving circuit, comprising the condenser and an inductance, may be either inductively coupled to the receiving antenna or directly coupled.

In any case, variable inductances are inserted in the antenna circuit and in the connected oscillation circuit, to bring the two into syntony with each other. If the coupling in the transmitting circuit is inductive, and also that in the receiving circuit, then there are four circuits which have to be brought into syntony with each other, viz. the closed sending circuit, the associated sending antenna, the receiving antenna, and the closed local receiving circuit, and unless this tuning is accurately done, the apparatus will be wanting in sensibility.

In practical radiotelegraphy the coherer apparatus with its complications of tapper, relay, and inker has now been almost entirely abandoned. The electrolytic detector is also not very widely used. Probably 90 per cent. of all the reception with spark systems is at present conducted with some form of contact rectifier or crystal rectifier, or else the author's glow-lamp oscillation valve detector, or with the Marconi magnetic detector. The latter has great advantages in military and naval work from its portability and simplicity.

If a carborundum crystal or zincite-chalcopyrite detector is placed in series with a high-resistance double receiver head telephone, we have the simplest and most easily adjusted form of detector of oscillations. The telephone *plus* the rectifier circuit is connected to the terminals of the adjustable condenser in the receiver circuit (see Fig. 19, Chapter VI.). When the spark happens at the distant sending station the train of electric waves which is sent out impinges upon the receiving antenna and sets up oscillations in the receiving circuit. The rectifier permits currents of electricity to pass better in one direction than the other, and hence the oscillations in the condenser circuit are to a considerable extent rectified into gushes of electricity in one direction through the telephone. The effect of repeated gushes corresponding in frequency to the spark frequency at the sending station is to create in the telphone a sound having the same frequency.

If the spark frequency is low, that is, about 50 to 100 per second, then the sounds in the telephone, long and short, corresponding to the Morse signals are difficult to distinguish from the sounds due to oscillations in the antenna produced by atmospheric discharges. If, however, the spark frequency is high, say, 300 to 500 or more per second, then the message-bearing signal sounds are shrill and musical, and can easily be distinguished from the atmospheric discharges or stray sounds.

The use of a high spark frequency has therefore become now very general, as by its use it becomes vastly easier for the receiving operator to distinguish the real signal sounds from the irregular, meaningless sounds produced in the telephone. This is the chief reason why alternators having a frequency of 300 to 500, or rotating dischargers such as Marconi's, have such advantages for radiotelegraphic spark system work.

So far, we have been considering only short distance receiving apparatus for spark telegraphy. If, however, an electric arc is used in the transmitting circuit, or other means for emitting from the sending antenna continuous or undamped waves, then some modifications are necessary in the receiving arrangements. If we suppose a receiving antenna connected in the usual manner through an oscillation transformer with a closed receiving circuit containing a condenser, as shown in Fig. 19 of Chapter VI., and if the terminals of the condenser are connected by a circuit containing a telephone receiver and some form of rectifier, such as a Fleming oscillation valve, or else a crystal detector, this arrangement works perfectly when the transmitter is sending out trains of damped electric waves. The damped oscillations then set up in the receiving circuit are caused to produce rectified gushes of electricity through the telephone receiver, one such gush corresponding to each spark at the transmitter. Hence, when the sparks follow in quick sequence, these brief currents of electricity coming at the same frequency through the telephone cause it to emit a sound of pitch corresponding to the spark frequency. This sound is cut up by the sending key into long and short sounds corresponding to the *dash* and *dot* of the Morse code.

If, however, the transmitter is sending out undamped waves then no such arrangement would work. The telephone would simply be traversed by a feeble unidirectional current which, although intermitted to form the signals, would create no corresponding sounds in the telephone.

The arrangements have, therefore, to be modified as follows:—

A telephone receiver has a condenser connected as a shunt across its terminals. The condenser shunted telephone is placed as a shunt across the terminals of a condenser of variable capacity which forms part of the closed receiver circuit. There is also inserted between the two condensers a make and break vibrating contact, called by Poulsen a *tikker*, which is vibrated by a separate electromagnet and serves to interrupt rapidly the connection between the shunted telephone and the main condenser in the receiving circuit (see Fig. 29). The action is as follows: When the tikker interrupts the connection of the shunted telephone and the main condenser, the undamped waves impinging on the receiving aerial set up persistent oscillations in the closed coupled condenser circuit, and create a potential difference between the terminals of the condenser. When the tikker next closes the circuit, the main condenser discharges into the condenser shunting the telephone, and this charge then leaks through the telephone in the form of a brief current. The telephone is thus traversed by intermittent currents which have the frequency of the tikker vibrations. If then the continuous waves are cut up into Morse signals at the sending end, the observer at the receiver hears these signals as long and short sounds in the telephone receiver.

The sounds have the frequency of the tikker vibrations, but as these are not quite regular, and, moreover, as the amplitude depends on the phase of the condenser charge at which the tikker closes the circuit, there is a want of uniformity.

A better and more sensitive receiver for undamped waves is the heterodyne telephone of Fessenden. If continuous waves fall on the receiving antenna, and are made to induce other oscillations in a closed receiving circuit in which a condenser has a crystal or other rectifying detector in series with a telephone as a shunt; then no signal sounds would be heard in the telephone, even if the continuous waves were cut up into Morse signals. The frequency is too great to overcome the inertia either of the telephone diaphragm or of the ear. Suppose that by some local high frequency alternator other continuous oscillations are induced in the antenna, the frequency differing slightly from those of the received waves. Then the superposition of these two sets of undamped oscillations will produce *beats* which have a frequency equal to the difference of the frequencies of the two undamped oscillations. This beat frequency can be adjusted to come within

S

the limits of good audibility.   Thus, for instance, let us suppose the incident undamped waves had a frequency of 50,000, and that by a local alternator we induce in the antenna other oscillations having a frequency of 49,500, the difference is 500, and the telephone would respond to this by sounding a shrill note.   When the key at the sending end cut up the incident into Morse signals, then the sounds in the receiving telephone would be correspondingly intermitted.

This method of reception also affords a means of cutting out stray atmospheric waves and false signals.

Goldschmidt has also invented an ingenious device for the reception of undamped waves from his alternator, called a *tone wheel*.   In this a metal wheel having teeth filled in with insulating material, is rotated at a speed such that for every complete alternation in the oscillation frequency, one tooth passes under a brush which makes contact with the wheel edge.   This wheel and brush is placed in series with a telephone and connected to the receiving antenna.   If the wheel is rotated a little faster or little slower than synchronous speed, a sound is heard in the telephone which can be cut up into Morse signals.

In connection with his methods of reception, V. Poulsen has also devised an ingenious photographic recorder.   The oscillations in the receiver circuits are made to affect a thermoelectric detector, and the current so produced is passed through a string galvanometer, consisting of a strong electromagnet having a single wire traversed by the current from the thermocouple placed in a very narrow air gap of the magnet.   Hence, even a very feeble current passed through this wire causes it to be displaced across the magnetic field, and in so doing it is caused to uncover a small slit in a plate and permit a ray of light to fall upon a moving strip of sensitive photographic film.   It thus records on the film a dot or a dash, according to the time during which the aperture is uncovered by the deflected wire.   The film, after being exposed, is immediately developed and fixed, and can be inspected after a few minutes.   Views of the recorder and photographic slip with signals on it are shown in Fig. 15.

5. **Systems of Intercommunication by Short Distance Radiotelegraphy.**—Owing to the fact that land telegraphy is in most countries a Government monopoly (the principal exception being the United States), and having regard to the special advantages of radiotelegraphy for supermarine communication, the great field of operations for it is found in communication between ships and between ships and the shore.

The Marconi Wireless Telegraph Company, Limited, formed in

1897 to work the Marconi system, began in 1899 to create such a system of intercommunication, and has since established all over the world a very large number of stations on the coast for communication with ships. The stations are established at very many places on the coast of Great Britain, Canada, the United States, Italy, and in other places, and vessels of numerous lines

110 words per minute 5 × 10⁻⁶ amp.

60 words per minute 1 × 10⁻⁶ amp.

[Reproduced from " The Electrician " by permission of the Proprietors.

FIG. 15.—Poulsen Photographic Receiver and Records.

working on the Atlantic Ocean are equipped with corresponding sending and receiving apparatus. A complete " wireless exchange " has thus been established, by means of which these vessels can communicate with each other when at sea, and with various ports.

The Marconi Company was not only the first in the field to establish, but even to-day (1915) is the only company operating a fully organized system of intercommunication with ships equipped with apparatus suitable for communicating with numerous shore stations. For this purpose certain wave lengths had to be selected; the wave lengths now ordered for intercommunication between ship' and ship and shore short distances are 300 metres and 600 metres. Each station and ship on which a wireless installation is made has its sending and receiving apparatus tuned for the same or for certain wave lengths. It is also designated by a letter or a pair of letters, called a "call-signal."

Thus, the Marconi wireless station established at the Lizard, in Cornwall, is denoted by L.D., and in the same way the Crookhaven station at the South of Ireland is denoted by C.K., and of the vessels on which installations are made, such as the ss. *Campania*, of the Cunard Company, the call signal is C.A., and

[*By permission of Marconi's Wireless Telegraph Co., Ltd.*]

Fig. 16.—Marconi Wireless Telegraph Cabin on ss. *Minnetonka*.

of the ss. *Deutschland* it is denoted by D.L. Some of these vessels are equipped with short distance apparatus: that is, for communicating with one another and with the shore up to 200 miles. Others are equipped with long distance apparatus for communicating with long distance stations, as described in a subsequent section. Each station established on the coast and each ship has, therefore, a certain range of operations, which, to some extent, depends upon the atmospheric conditions.

On board the vessels a special telegraph cabin (see Fig. 16) is set apart for the radiotelegraphic work, and, equipped with the apparatus for sending and receiving, as already described, worked by a skilled operator. A view of the interior of one of a ship's

Marconi cabin is shown in Fig. 17. The transmitting apparatus is seen on the right hand and the receiving apparatus on the left, and the Morse printer and sending key in the centre. On the ship the antenna is suspended from a gaff attached to the main or fore mast, and brought in through an insulator to the cabin. The Marconi shore stations are, by special agreement with the British postal telegraphic service, connected with the land lines

FIG. 17.—View of the Interior of a Ship's Marconi Cabin with Wireless Telegraph Plant.

of the country, so that they are in close correspondence with every place in which there is a postal telegraph office.

Supposing, then, that the ss. *Campania*, approaching the south of Ireland, desires to communicate with Crookhaven station in the south of Ireland. Her operator sends the call signal C.K. several times at intervals, waiting in between to see if there is any response, and on getting the return call signal from Crookhaven, he establishes communication by stating the name of the vessel, its approximate position, and course. Then follows the message which the ship desires to send, and the shore station would acknowledge receipt of same, and perhaps repeat it for safety. If the message is for a private person, it would then be despatched from the coast station over the postal telegraph lines to its destination. In the same way, a message sent to a coast station for a particular ship is communicated to that ship. The operators in the coast stations are provided with charts showing the times of sailing from the various ports of all the vessels equipped with the wireless telegraph apparatus belonging, say, to the Marconi Wireless Telegraph Company. By consulting this chart, the operator can tell what vessels are at any time within range of his station. He can then call up some particular vessel by signal and establish communication with it if it happens to be within range. If, however, the desired vessel is not within range, but another vessel also equipped with wireless apparatus of the same system is within reach, the message can be despatched to the vessel within reach, which is then requested to forward it to the vessel lying beyond, within reach of itself but out of reach of the shore station. Thus, for instance, the operator at Crookhaven might have a message for the ss. *Campania* when four hundred miles out at sea, and not being able to reach the vessel at this distance, might despatch it to another vessel, say ss. *Lucania*, which is at a distance of a couple of hundred miles, and request the ss. *Lucania* to fling the message forward on to the *Campania*. In this manner a message may be made to jump over three or four ships, arriving at its destination after two or three retransmissions. This, of course, is on the assumption that the vessel is not provided with long-distance receiving apparatus, for if it is so provided it can be reached directly by long-distance stations, to be described presently.

The difficulties with which radiotelegraphy of this kind has to contend arise chiefly from atmospheric electric discharges, from possible interferences or cross conversation, and very occasionally from deliberate interference. It has already been explained that a transmitting and receiving apparatus can be made syntonic by proper tuning of the circuits, and will therefore become receptive only of signals approximately agreeing in wave length with the frequency for which the stations are tuned. Thus, for instance, long-distance stations, to be presently described, operate generally with a long wave which does not in the least degree affect apparatus used on board ships and shore stations, operating with a wave length, say, of 1000 feet. On the other hand, receiving stations tuned for 1000 feet will be receptive for waves of that length, and also for others not differing very greatly therefrom. The difference in wave length which can exist between a station's own proper wave length and that of the incident waves without stopping reception is a somewhat variable quantity, and depends essentially upon the damping or decrement of the sending station, and also the damping or decrement of the receiving station, that is, upon the form of the resonance curve of the two stations taken together. This resonance curve, as already shown, is more peaked or sharper in proportion as the decrements of the sending or receiving station are smaller. Hence, if the sending station is emitting undamped waves, a more exact tuning or syntony will be required on the part of the receiving station in correspondence to obtain the best effect, than if the transmitter is sending out damped wave trains. That which is really inimical to the privacy of communication is the emission by various transmitting stations of powerful highly damped waves with large initial amplitude, whilst greater privacy is secured by the emission of feeble undamped or very slightly damped trains of waves. Hence the employment of strong damped waves should be as much as possible repressed in the general interest.

In the same manner the receiving station will secure its own privacy far better when its receiving circuits are circuits only slightly damped and at the same time largely inductive, because then a slight difference between the frequency for which the receiving circuit is tuned and the frequency of the incident waves will reduce the amplitude of the oscillations set up in the receiving circuit by a very large amount, and hence enable the receiving operator to cut out those signals he does not desire to receive by exact tuning with the wave length of those he does desire to receive. Hence what may be called the unintentional interference and the picking up of messages it is not desired to receive is to some extent a matter of organisation, to a large extent a matter of apparatus, and also of personal skill on the part of the receiving operator, and if he possesses the requisite means and apparatus, he can render himself, so to speak, deaf to everything except the æthereal vibrations to which he desires to be sensitive. On the other hand, there are certain disturbances which are produced by atmospheric electric discharges which create so-called vagrant waves, and these, if sufficiently powerful, affect even syntonic apparatus. They are technically termed atmospheric X's. These atmospheric disturbances are particularly marked in tropical regions at certain times of the day and year. They exhibit themselves in the case of the telegraphic recording receivers by making dots and dashes irregularly upon the tape, which are mixed up with the dots and dashes belonging to the Morse signals being received, and render them more or less unintelligible. In the same way, when using the telephonic receivers, the atmospheric X's cause sounds of irregular duration and magnitude sufficient to confuse the signals. These atmospheric disturbances have been particularly studied of late years.

Many valuable observations have been put on record by Admiral Sir Henry Jackson,[1] and further reference to them is made in a later section of this chapter.

A great many devices have been described and patented from eliminating these atmospheric disturbances. In the case of damped wave-transmitters, the best protection is the use of a high frequency regular or musical spark. The operator then easily distinguishes between the message-bearing and the stray signals.

All protective devices depending on resonance act very imperfectly, for the reason that when an irregular or vagrant wave strikes a receiving antenna, it sets it in electrical vibration in accordance with the natural frequency for which that aerial is tuned, and therefore affects the receiver. The signals coming from a properly tuned corresponding transmitter do the same, and the only difference is, that the vagrant wave being highly damped, makes less effect than the proper sending station. Nevertheless, no protective device depending on resonance is perfectly eliminating.

In regard to radiotelegraphy, these atmospheric disturbances occupy the same position that earth currents and magnetic disturbances do towards telegraphy with wires. It is well known that at certain times the earth's magnetism is in a state of great disturbance called a magnetic storm. Periods of most frequent magnetic storms approximately coincide with the periods of most frequent sun spots and most frequent auroræ, and at the time of these magnetic storms electric currents circulate in the crust of the earth which sometimes interrupt ordinary telegraphic communication by wires altogether.

6. **Long Distance Radiotelegraphy.**—The chief difference between radiotelegraphic stations established for short distance work, say, up to 200 miles or so, and those required for communication over 1000 miles or more, is in the increased power and wave length required for the longer distance. This requires certain modifications, chiefly in the transmitting apparatus, with the

---

[1] For details of these observations the reader may either consult the original Paper of Admiral Jackson (see *Proc. Roy. Soc. Lond.*, vol. 70, p. 254), 1902, or a summary in the author's book, " The Principles of Electric Wave Telegraphy," 3rd Edit., Chap. X. (Longmans & Co.).

object of producing long electromagnetic waves of great amplitude. In power stations emitting damped waves the methods employed for production on a large scale are in principle the same as in small stations, but the apparatus has to be suitably modified. In place, therefore, of induction coils operated by batteries, we have to employ alternating current transformers to charge large condensers and alternating current dynamos to supply these transformers. Hence, a radiotelegraphic power station comprises, in the first place, a source of motive power, which may be a steam or oil engine. Steam is in every way preferable where water can be obtained, but in isolated places an oil engine is a necessity. The engine may be coupled directly to an alternator or may drive it by a belt, the alternator generally being of a type known as a revolving field fixed armature alternator, and having an electromotive force, say, of 2000 volts, and a frequency of 300 to 500. The current from this alternator is supplied to a battery of high tension transformers which may have their primary coils joined in parallel and their secondary coils joined in series. These transformers should be oil insulated. The transformers are connected through certain choking coils or inductances with a battery of condensers, which may be glass plate or tube condensers, which in turn are connected in series with one coil of an oscillation transformer, the secondary circuit of which is inserted between an antenna and the earth, and with a spark gap which is across the terminals of the transformers. In some cases, the antenna is directly connected to the condenser circuit. Owing to the large quantity of electricity which passes at each discharge, it is desirable that this spark discharger should be of the revolving ball or disc type, such as that devised by the author, in which the spark balls are slowly revolved by means of electric motors, so as to continually expose fresh surfaces, or else the greatly improved high-speed-disc dischargers invented by Mr. Marconi. The currents are controlled and the signals made by short circuiting an inductance coil, which may be included in the primary circuit of the transformers. The oscillations transformer or inductance in circuit with the condenser is generally oil insulated. One of the most important elements of a power station of this description is the antenna. The creation of a long wave necessitates a correspondingly elaborate structure as an antenna. This antenna, as in the Marconi power stations at Poldhu in Cornwall, and Cape Breton, Nova Scotia, were at first supported by wooden towers, 215 feet or more in height, and 25 or 30 feet square at the base, but are now upheld by steel tubular masts. These masts sustain a Marconi directive antenna of the form shown in Fig. 3, partly horizontal and partly vertical, the horizontal portion being sustained by other towers or masts (see Figs. 2, 3, and 4, above).

The first long distance radiotelegraphic power station in the world was that undertaken in 1900, on the decision of Mr. Marconi and the Marconi Wireless Telegraph Company to make a serious attempt to achieve by his methods radiotelegraphy across the Atlantic. A site was accordingly obtained in August, 1900, at Poldhu, near Mullion, in Cornwall, at a place far removed from large towns, and where work could be conducted with privacy. Mr. Marconi designed a multiple wire antenna of a cone shape to be supported by a ring of masts 200 feet in height, the station being placed in the centre. Work was commenced in October of the same year on this great enterprise.

The author was entrusted with the duty of designing and arranging the machinery in this first power station for creating the powerful electrical oscillations necessary to excite oscillations in a large antenna. As the site selected for the station was on a cliff some way from water, an oil engine (25 h.-p.) was employed as a prime mover, and this drove by means of a belt, an alternator giving an alternating current of 50 periods and an E.M.F. of 2000 volts. This voltage was raised by transformers to 20,000, and employed to charge glass plate condensers.

In the early experiments the author arranged a double transformation system, in which the secondary terminals of the transformer were connected to a pair of spark balls, and these were also connected by a condenser in series with the primary circuit of an oscillation transformer (see Fig. 18). The secondary circuit of this oscillation transformer was again connected with a pair of spark balls, and these again with a second condenser and primary of an oscillation transformer, the secondary circuit of which was inserted between the antenna and the earth. Between the alternator, D, and the transformer, T^1, was inserted a pair of choking coils, H^1, H^2, by the short circuiting of which the signals were created. This plant was completed in August or September, 1901.

In the earliest experiments the author designed a fixed ball spark gap, but it was found that the heavy discharges wore away the balls rapidly. He then designed a form of spark gap consisting of two heavy metal discs like solid wheels with thick rims. These were rotated by electric motors and the oscillatory discharge was caused to take place between the rotating discs.

In course of time, however, these early arrangements were considerably modified, but they sufficed to create sufficiently powerful oscillations to produce electromagnetic waves which were detectable across the Atlantic. After much experimenting over shorter distances, Mr. Marconi went, in December, 1901, to Newfoundland, taking with him balloons and kites as means for raising a wire to form a temporary receiving antenna and various forms of oscillation detector; and on December 14, 1901, he was able to announce that he had received signals which undoubtedly were the pre-concerted signals at that time being sent out from the Poldhu station. It was thus demonstrated that the electromagnetic waves made by no extravagant power expenditure in Cornwall could be detected at a distance of nearly 2000 miles, in spite of the considerable curvature of the sea surface in that distance.

This achievement sufficed to give encouragement to Mr. Marconi and his supporters to proceed with the enterprise with the object of establishing regular commercial radiotelegraphic communication across the Atlantic.

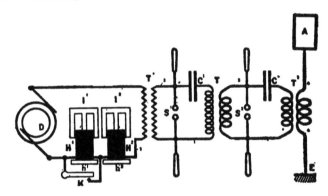

Fig. 18.—Apparatus for Multiple Transformation of Oscillations (Fleming).

Marconi returned to England in February, 1902, and at once made arrangements for the erection at Poldhu of a permanent structure for carrying a large antenna. This consisted of four wooden lattice towers, 215 feet in height, placed at the corners of a square 200 feet in side. These towers were strongly stayed by steel wire ropes. At first a conical antenna was employed, but later on, after Marconi had invented the bent antenna, other masts were erected to carry an antenna partly horizontal and partly vertical. New buildings for the generating plant were erected in the centre of the square, and more powerful machinery employed. and improvements introduced which experience had indicated. At the same time similar stations were erected at Cape Cod, in Massachusetts, U.S.A., and Cape Breton, in Nova Scotia.

Whilst these improvements were in progress, Marconi returned to Canada, and on the way across conducted interesting experiments on board the Atlantic liner ss. *Philadelphia.* An insulated antenna wire, 60 metres high, was fixed to the ship's masts. Messages sent from Poldhu were received on board as the vessel went west, and printed down on the Morse tape. Readable messages were obtained in this way up to 1551 miles from Cornwall, and communications or signals up to 2099 miles, by means of Marconi's printing telegraphic apparatus.

In July, 1902, he conducted similar experiments on board the Italian warship *Carlo Alberto,* placed at his disposal for this purpose by the Italian Government, and on this occasion he employed his magnetic detector as a receiving instrument (see Chapter VI.), the invention of which he had patented some time previously. The first voyage of the *Carlo Alberto* was to the Baltic, and messages were received on board from Poldhu as far as Cape Skagen in Denmark and Cronstadt in Russia.

In August, 1902, the *Carlo Alberto* proceeded to the Mediterranean, and continued to receive wireless messages all the way; and later on in the same year went across the Atlantic to Nova Scotia, receiving messages from Poldhu during the voyage, and whilst the ship was lying in Sydney Harbour.

Towards the end of 1902 the stations being erected in Nova Scotia and at Cape Cod were sufficiently advanced to enable a preliminary test to be undertaken, and on January 19, 1903, a wireless message was transmitted across the Atlantic from Welfleet, Cape Cod, Massachusetts, U.S.A., to Poldhu in Cornwall, from President Roosevelt to King Edward VII., whilst at the same time other messages were sent from Cape Breton in Nova Scotia to Poldhu, and a large number of messages were transmitted both ways across the Atlantic in that and the following year.

In 1904, Marconi established a system of long distance wireless telegraphy, communicating news to the principal Atlantic liners in the course of their voyages from Liverpool to New York, small newspapers being published on board containing news paragraphs transmitted by radiotelegraphy from the mainland on both sides, and by 1905 this system had become established as an indispensable means of communication with vessels *en voyage* across the Atlantic.

In order to conduct this system of communication between ships without interfering with the transatlantic work, new stations were erected on both sides of the Atlantic by the Marconi Company—one at Clifden in Ireland, and the other at Cape Breton in Nova Scotia—in which many improvements were

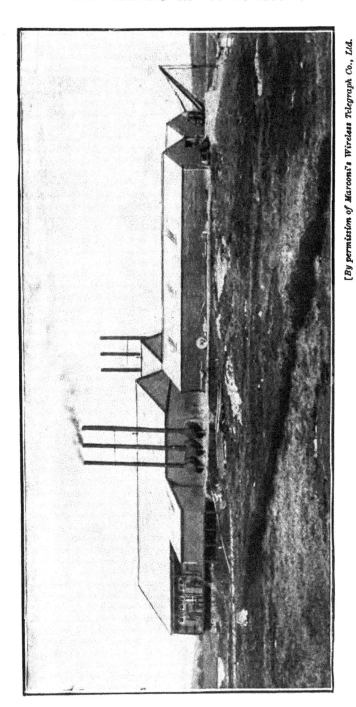

[*By permission of Marconi's Wireless Telegraph Co., Ltd.*

Fig. 19.—View of the Engine and Boiler-house of Marconi's Transatlantic Radiotelegraphic Station at Clifden, Ireland.

[*By permission of Marconi's Wireless Telegraph Co., Ltd.*

FIG. 20.—View of the Condenser-house and Antenna of Marconi's Transatlantic Radiotelegraphic Station at Clifden, Ireland.

introduced by Mr. Marconi, having for their object the increase in speed and certainty of sending and receiving. This station was completed in May, 1907, and by October, 1907, he was able to commence a regular system of press message radiotelegraphy across the Atlantic, communicating news for the American daily journals, and also exchanging public and private messages (see Figs. 19 and 20).

The Marconi Company have at the present time (1915) six high power transatlantic stations for radiotelegraphy, which are equipped on the spark system, namely, those at Clifden in Ireland, Poldhu in Cornwall, Cape Cod in Massachusetts, U.S.A., Glace Bay in Nova Scotia, and at Carnarvon in Wales, and New Jersey, U.S.A.

The station at Poldhu has been chiefly used for long distance communication with Atlantic liners, a large number of which are equipped with long distance receiving apparatus, whilst the other stations are reserved for the transatlantic communication proper.

As at one time statements were made that the working of these large stations would interfere with the ordinary ship to shore telegraphy, special demonstrations were arranged for the purpose of disproving these statements by Mr. Marconi; and these took place under the inspection of the author on March 18, 1903, in which a large number of messages were simultaneously sent out from the Poldhu station, and also at the same time messages were sent out by short distance apparatus from a small station close to the power station. These were simultaneously received by Mr. Marconi at the Lizard station six miles away, and were also received by another observer at a station near Poole, two hundred miles away; and it was demonstrated that the powerful waves sent out by the long distance apparatus in no way whatever interfered with the messages being sent by the short distance apparatus in close contiguity to the power station.

These tests were confirmed some months later by Admiralty officials, when similar demonstrations were made between Cornwall and Gibraltar.

Returning, then, to the details of the above power stations, we may notice the highly effective form of discharger invented by Marconi for operating with large discharges.

It has already been pointed out that in power stations on the spark system it is essential that the discharge across the spark balls should be a dead-beat discharge of extremely uniform frequency, and should consist wholly of electricity which has been drawn from the condenser, and not be mixed up with a true alternating current arc discharge which is due to current coming

directly out of the transformer. Marconi, however, found by
careful experiments that between metal surfaces in exceedingly
rapid relative motion it is very difficult to produce a true electric
arc, but, nevertheless, an oscillatory discharge from a condenser
can pass between these surfaces. Hence he has devised forms of
high speed rotating discharger made as follows :—

In one form there are a pair of metal discs, C, C (see Fig. 21),
which are caused to rotate rapidly by electric motors or other
means. Between these, and insulated from them, another disc, A,
rotates at a high speed, with its plane at right angles to the
other two. The terminals of a dynamo machine are connected
with the terminals of a pair of condensers, K, in series, and

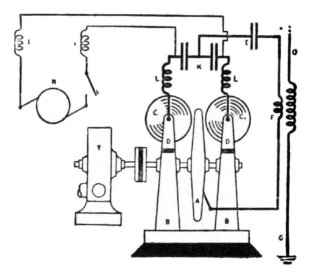

Fig. 21.—Marconi High-speed Disc Discharger.

through inductances, L, L, with the discs C, C. The middle point
of the condensers is connected to another condenser, E, and this
through one coil of the jigger F with disc A. If a key in the
dynamo circuit is closed, it charges the condensers, K, and then
at a certain potential the condenser E discharges with oscillations
across one or other of the air gaps between the rapidly revolving
wheels. The arc discharge which attempts to follow in the track
of the oscillations is, however, prevented by the rotation of the
discs from taking place.

In another arrangement (see Fig. 22) the disc A has studs on
its circumference at intervals placed transversely to its plane.

The action of the discharger as shown in Fig. 21 is as follows :

The dynamo charges the two large condensers, K, and then at some potential that plate of the small condenser E which is in connection with the disc A is charged by a spark passing between the central disc and one of the side discs. Suppose this action charges the said plate of the small condenser positively. This then strengthens the electric field between the middle disc and the other side disc, and a discharge happens on that side which reverses the sign of the charge on the small condenser plate attached to the middle disc. These reversals of charge rapidly succeed each other, each taking place with oscillations, and the

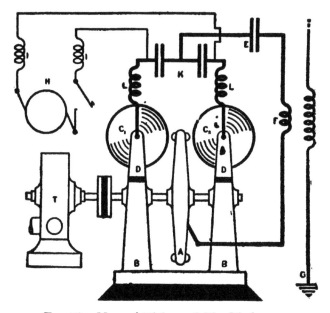

Fig. 22.—Marconi High-speed Disc Discharger.

effect is to produce almost unintermittent oscillations in the coil F, in series with the small condenser E. This discharger, therefore, affords a means of obtaining practically unintermittent oscillations from a continuous current dynamo machine, whilst the rapid rotation of the discs prevents the formation of a direct current arc which would otherwise stop the process. The principle of the discharger is therefore entirely different from that of the Duddell-Poulsen method of obtaining undamped oscillations from an electric arc. In the case of the discharger, as shown in Fig. 22, the discharges are intermittent, but succeed each other very rapidly.

T

These Marconi dischargers work with great efficiency, and permit very rapid signalling to be conducted with them.

The new Cape Breton or Glace Bay station in Nova Scotia has an antenna of 200 wires rising 220 feet vertically, and then extended 1000 feet horizontally, at a height of 180 feet above the ground. The antenna was designed for a wave length of 20,000 feet. The condenser used has a capacity of 1·8 microfarad, and the spark length used is generally 18 to 20 mm., equivalent to a charging voltage of nearly 46,000 volts.

The bent antennæ at Glace Bay, Nova Scotia, and at Clifden in Ireland, were placed with their free ends pointing directly away from each other.

At the end of May, 1907, the Clifden station was completed. It was designed also for a wave length of 20,000 feet, and had a condenser of 1·8 microfarad, charged to 46,000 volts. This condenser is an air condenser formed of sheets of metal hung up on insulators, thereby avoiding the dissipation of energy in condensers made with glass dielectric.

The dischargers used as spark gaps in these stations are of the revolving disc type above described, and permit of signalling at a rate as high as hand sending can accomplish. Owing to the regularity of the discharges, the Morse dash is heard in the telephone at the other side as a clear musical note, and the operator can distinguish between it and the irregular sounds due to atmospheric discharges.

These large stations at Cape Breton and Clifden began exchange of radiotelegraphic messages across the Alantic on October 17, 1907, and hundreds of thousands of words in Press and private messages have since been transmitted, whilst communication has never been interrupted by causes within the radiotelegraphic stations themselves, and only for seventeen hours by breakdowns on land lines. The long history of this great achievement was related by Mr. Marconi in a lecture at the Royal Institution of Great Britain, delivered on March 13, 1908, in which he recounted the various stages of the work and the steps by which success had finally been attained.

The practicability of long distance radiotelegraphy having been thus demonstrated by Mr. Marconi, numerous rivals entered the field with projects for transatlantic telegraphy, and in the case of the stations erected on the spark system, the general arrangements first introduced in the Marconi stations have been closely followed, with the exception that the coupling of the antenna in these other stations is generally a direct coupling instead of inductive.

A few details of the largest and latest Transatlantic Marconi Station at Carnarvon may here be given. This station is constructed for duplex working, that is, for sending and receiving simultaneously with a corresponding station in the United States

FIG. 28.—View of 800 K.W. Main Motor-alternators in Carnarvon Marconi Transmitting Station.

at New Jersey. The receiving station on each side is therefore placed at some distance from the sending station. In North Wales the sending station is on the side of a hill near Carnarvon, and the receiving station is at Towyn, 60 miles away on the

Welsh coast. Four wires connect these stations, by means of which the transmitters at Carnarvon can be worked from Towyn. Power is supplied to the sending station by an overhead line at 10,000 volts. This comes from the North Wales Power and Traction Company, and is a three-phase alternating current supply. The power supply station is at Cwn Dyli, 11·5 miles away, and obtains its power from water supplied by a lake on Snowden. The 10,000 volt three-phase current is reduced by transformers to 440 volts and employed to drive three-phase motors coupled to single-phase alternators or to direct-current generators as required. The main transmitting sets consist of 300 K.V.A. single-phase alternators supplying current at 1750 volts and 150 frequency (see Fig. 23).

The Marconi studded disc discharger is coupled direct to the shaft of the alternator. The alternator sends its current to a bank of 75 K.V.A. transformers, which step up the pressure, and these charge a condenser consisting of metal plates placed in oil. These condensers are discharged by the rotating discharger through the primary circuit of an oscillation transformer, the secondary of which is in series with the antenna. This antenna is a directional antenna approximately 3600 feet long and 500 feet in width. It is supported on ten tubular steel masts 400 feet high (see Fig. 24). These masts are stayed by wire ropes and rest on immense blocks of concrete. The earth-plate is a very extensive system of copper plates buried in the earth. The signalling switches are relay switches which can be operated from the receiving station 60 miles away, or else by hand in the transmitting station.

The receiving station is provided with two antennæ. One of these is the main antenna by which signals are received from the other side of the Atlantic. To prevent this receiving antenna from being affected by signals sent out from the Carnarvon trans-mitting station the receiving station has a second smaller hori-zontal antenna called the balancing antenna. The effect of the waves from the Carnarvon transmitting station striking on these two receiving antennæ is to produce in them electrical oscillations which can be made to neutralise one another on a receiving circuit which is coupled to both the antennæ. The waves from the distant transmitting station across the Atlantic are, however, only able sensibly to affect the high receiving antenna. Hence, whilst the receiving station is sensitive to the waves arriving from the long distance, it is insensitive to the waves arriving from the adjacent sending station. Communication can therefore go on both ways simultaneously across the Atlantic without hindrance. This just doubles the available transmitting time.

It remains to be added that the transmission can be carried on by automatic methods, in which the message is first perforated in code on paper tape. This tape is fed through an automatic transmitter which operates relays, and these again operate the signalling

Fig. 24.—Part of Carnarvon Wireless Telegraph Station (Marconi), showing Entrance Point of Antenna Wires.

keys and control the spark discharges. This control can be effected from the receiving station, which is 60 miles from the transmitting station. In the same manner the received signals can be made to operate an Einthoven galvanometer and to photograph themselves on sensitive paper tape.

A view of part of the Carnarvon station buildings, showing the entrance of the aerial wires, is given in Fig. 24, and a view of one of the 300 K.W. main motor-alternators in Fig. 22.

As an example of a station combining in one the spark and the arc methods, we may take that at Cullercoats, on the Northumberland coast, about 8 miles from Newcastle, England, originally erected by the Amalgamated Radiotelegraphic Company

[*Reproduced from " The Electrician " by permission of the Proprietors.*

FIG. 25.

but since taken over by the General Post Office and reconstructed. The station itself is situated on a promontory running out to sea. It comprised a small four-roomed one-storied building and a large umbrella antenna supported by a single wooden lattice tower (see Fig. 25). The mast was built up of bulks of timber 6 inches square, jointed in lengths. It is 220 feet high and 2 feet square at the base, and supported in a foundation of concrete and stayed by wire ropes cut up into lengths by insulators of creosoted wood. The antenna was constructed of bronze wires (see Fig. 25), which extended from the top of the mast and spread over a circle of 220 feet in diameter. It was made in two parts, each consisting of 12 wires, and stretched out into a wide semicircle by guy ropes attached to anchors fastened to various rocks. The 24 wires of the complete antenna were connected at their lower ends to one wire which encircles the mast at a height of about 100 feet. The upper ends of the two halves were connected to two cables which came down into the station building. It will thus be seen that if the two cables are not connected together, the wires form a loop antenna, but if they are connected together they form a single antenna. The earthplate at the station consists of a large number of wires buried about 2 feet in the ground, radiating in all directions from a point near the foot of the mast. The station contained both spark and arc apparatus. In the case of the spark apparatus the power was supplied by an 8-h.p. motor driven directly from the town electric supply, and was coupled to a 5-kw. alternator, supplying 14 amperes at 400 volts, and at a frequency of 120. This alternating current was raised by a transformer to 50,000 volts. The sending operator could start and stop the alternator by a switch from the operating table. The primary current of the transformer passed through a sending key, to interrupt it in accordance with the signals of the Morse alphabet. The high-tension alternating current was led to a third

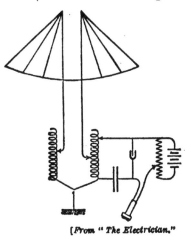

[*From " The Electrician."*]

FIG. 26.

room, in which there is a large battery of Leyden jars and an associated inductance and spark gap, forming the oscillatory circuit. This inductance was directly connected to the antenna through a large switch which changed over the antenna from the transmitting to the receiving system when receiving. There were the usual tuning coils inserted in the circuit of the antenna. The reception was conducted by means of an electrolytic oscillation detector and a telephone, the antenna being arranged as a loop antenna (see Fig. 26). With this apparatus communication was carried on for about 400 miles between Cullercoats and Christiana.

The station also contained a Poulsen arc apparatus. The arc generator consisted of a metal box with marble ends, shown at the

left-hand bottom corner of Fig. 27. This box contained the copper
and carbon electrodes, the cooling of the copper anode and of the
arc box being effected by radiating flanges exposed to the air, and
not by water circulation (see Fig. 9, § 3). The striking of the arc
was accomplished by lifting the copper electrode momentarily by
a lever, and then allowing it to fall to an adjusted distance. The
box was kept full of hydrogen supplied from a gas cylinder, or
from a calcium hydride generator, by which hydrogen is generated
by dropping calcium hydride into water. About two pounds of

[*Reproduced from " The Electrician" by permission of the Proprietors.*

Fig. 27.—Poulsen Arc Apparatus in Cullercoats Station.

hydride provide enough hydrogen for 60 hours' continuous work.
In some cases coal gas is used instead of hydrogen.

The carbon cathode is rotated by clockwork. The usual telegraphic work was carried on with a single copper-carbon arc having
a fine adjustment for arc length, the arc being formed in a strong
magnetic field perpendicular to it. The windings of the electromagnet are in series with the arc as well as with a variable resistance, and the arc is formed by a continuous current of 480 volts
taking 10 or 12 amperes. The oscillation circuit is arranged as a
shunt to the arc with a direct connection to the antenna, as shown

in Fig. 28. It comprises an inductance coil of many turns and a condenser formed of zinc plates immersed in oil. The plates are separated by a distance of 3 mms., and the capacity is arranged in two sections, so that although a point on the inductance coil is put to earth, the terminals of the arc remain insulated. A variable condenser is connected in parallel with the fixed condenser to enable changes to be made in the emitted wave length, which is usually between 1200 and 1500 metres. A hot wire ammeter is inserted in the earth connection to show the current passing into the antenna.

The signalling is effected by short circuiting a few turns of the inductance coil, and therefore altering the wave length of the emitted waves. The frequency employed is about 200,000, and the current in the antenna about 10 amperes. The receiving

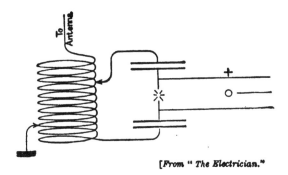

[*From " The Electrician."*

FIG. 28.

apparatus used with these undamped waves consists of an oscillation transformer of which the two circuits are very loosely coupled, the primary being joined to the terminals of a condenser inserted in the antenna circuit, and the secondary connected to another large condenser, and also intermittently to a telephone shunted by a third condenser (see Fig. 29). The connection between the telephone and the condenser circuit is made by means of a *tikker*, or vibrating electromagnet, in which a very light rapidly moving hammer closes and opens the circuit. When the circuit is open, energy accumulates in the large condenser, and on closing it some of it passes into the condenser of the circuit, and on the opening of the contact again this condenser discharges through the telephone. The contact points of the ticker are made of crossed gold wires, and the vibrating mechanism is enclosed in a small sound-proof box. The observer, therefore, hears as sounds of longer or shorter duration in the telephone the more or less prolonged short-circuiting by the sending key of part of the inductance in the transmitting circuit. Owing to the loose coupling, the tuning is very sharp, and it is easy to perceive in this receiver the effect of altering about one-half per cent. in the capacity of the sending circuit

The advantages claimed for the arc method of signalling are first its silence, and, secondly, the entire absence of sparking at the sending key, also the greater compactness of the apparatus and the lower voltages dealt with. For example, the maximum potentials which occur at the top of the antenna when the undamped waves are being used are probably not greater than two or three thousand volts, and the insulation required in the apparatus itself is only for voltages of the order of 1000 volts. It

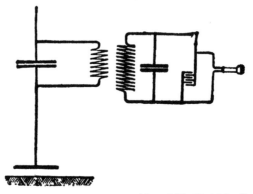

[From " The Electrician."

FIG. 29.

is also affirmed that atmospheric disturbances are much less felt when using the undamped wave apparatus than when using the damped waves. Furthermore, it is claimed that comparative tests of the arc and spark methods, carried out over ranges of about 900 miles, have shown that the undamped waves are less obstructed by mountainous country than are the damped waves of a spark transmitter of the same wave length when using about the same sending power. It remains to be seen, however, whether in actual working these differences will give rise to a marked advantage.

The Amalgamated Radiotelegraphic Company began to erect another large power station in Ireland, near Knockroe farm, thirteen miles from Tralee, co. Kerry, with a view to correspondence with another station to be erected on the coast of Nova Scotia. The station was to consist of a power house 40 feet square, with an accumulator house and two operating rooms at a distance from the power house. Three high masts, each about 3 feet square at the base and 360 feet high, are arranged round the operating house, and nine short masts, each 70 feet high, form a circle 2000 feet in diameter round the high masts. The high masts are built of square timber baulks framed together in a nearly horizontal position on a staging, and then raised into a vertical position by a

[Reproduced from " The Electrician " by permission of the Proprietors.

FIG. 80.—The Poulsen Receiving Apparatus in Cullercoats Station.

crab and jury mast. These high masts carry the upper insulated ends of an antenna of about 300 wires which descend in a cone to the lower masts, whence they are gathered into one conductor and led to the operating house. The power house contains two large and two small dynamos driven by a portable steam engine. The larger machines are intended to supply the continuous current at 500 volts for a Poulsen arc.

The condenser included in the oscillation circuit consists of metal plates hanging in air of a total capacity of $\frac{1}{30}$ of a micro-farad, and capable of being arranged in various ways. With this apparatus it is estimated that a radiation of 10 to 15 kilowatts will be reached, employing a wave 3000 to 5000 metres in length.

7. **The Effect of Atmospheric Conditions and Terrestrial Obstacles on Long Distance Radiotelegraphy.**—The first attempt to conduct radiotelegraphy extending over many hundreds of miles revealed the important influence which atmospheric conditions have upon such telegraphy, and especially the effect of sunlight upon it.

In one of his voyages across the Atlantic, when receiving signals on board ss. *Philadelphia*, Marconi noticed that the signals were received by night when they could not be detected by day. He arranged a programme of experiments for sending signals of given strength on certain days from the power station at Poldhu, from 12 to 1 a.m., 6 to 7 a.m., 12 to 1 p.m., and 6 to 7 p.m. Greenwich mean time, every day for a week. He found that on board the *Philadelphia* he did not notice any apparent difference between the signals received by day and by night until the vessel had reached a distance of 500 miles from Poldhu. At about 700 miles, signals transmitted during the day began to weaken, while those received at night remained quite strong up to 1551 miles, and were even quite decipherable up to 2100 miles from Poldhu, being recorded with his coherer and printing receiver. He noticed, also, that the weakening occurred at the time when daylight first fell upon the transmitting antenna, and he inferred that the cause of it was the dissipating action of light upon negative charges of electricity. Another explanation was suggested soon after. It is well known that the atmosphere, especially at high levels and in sun-light, is in a state of ionisation, and it has been shown by Prof. Sir J. J. Thomson that these gaseous ions or point charges of negative and positive electricity are set in motion by a long electric wave travelling through space, and they therefore partially absorb the wave energy. By means of an apparatus devised by Ebert and Gerdien, it is possible to measure the conductivity of the atmosphere in any state and so to determine the number of ions or electrons in a cubic centimetre. Experiment shows that the numbers of positive and negative ions are considerable, but generally unequal. Using an Ebert apparatus, Boltzmann had found during an Atlantic voyage from Dover to New York that the number of electric ions present in the Atlantic atmosphere was 1150 positive and 800 negative per cubic centimetre. During a voyage from Montreal to Liverpool, A. S. Eve found from 600 to 1400 positive and from 500 to 1000 negative per cubic centimetre, the ratio of positive to negative varying from 1·04 to 1·83. These numbers do not greatly differ from those found over land areas of large dimensions, such as Germany or Canada.

From data given by Strutt as to the amount of radium in sea-water and in various sedimentary and aqueous rocks, Eve draws the conclusion that this ionisation cannot be wholly accounted for by radioactivity of the sea or soil. Knowing, however, that ultra-violet light is a cause of ionisation, and, perhaps, the penetration of the upper layers of the atmosphere by cosmical matter carrying electric charges is another possible cause, we may no doubt assign to these agencies some share at least in the production of atmospheric ionisation. In any case, it is clear that the terrestrial atmosphere, when we are concerned with large volumes of it, and especially on that side of the earth which faces the sun, cannot be considered as equivalent to space occupied merely by free ether, or even air and ether.

The presence of these ions or electrons in the atmosphere imparts electric conductivity to the air, and hence it was assumed that they would exercise an absorptive action upon the energy of long electric waves. In other words, air exposed to sunshine, although it may be extremely transparent to light waves, might perhaps act as a slightly turbid medium for long electric waves; but the effect is not sensible up to distances of two or three hundred miles. The atmosphere, moreover, when in a state of ionisation exercises a certain selective absorption upon long electric waves, just as various transparent media exercise selective absorption upon light waves of certain wave lengths. A wave of long wave length and small amplitude is less obstructed than one of lesser wave length and larger amplitude. Accordingly, by the choice of a suitable wave length and amplitude, waves can be generated which are not much subject to absorption by the daylit atmosphere, and considerable progress has been made of late in a knowledge of the particular wave lengths to employ for long distance radiotelegraphy.

From the known conductivity of air at such heights as are accessible to us it can be shown that the above described form of absorption of wave energy by ions in the atmosphere would not account for the observed effects on radiotelegraphy, and that the diminution of signal strength due to daylight is not to be explained by a true absorption.

Hence another theory called the theory of ionic refraction has been proposed. It has been shown by Dr. Eccles that an electric wave will travel slightly faster in an atmosphere populated with heavy electric ions, and hence that if such ionisation takes place at a high level in the atmosphere, electric rays sent up to it may be refracted down again by an action similar to the bending of rays of light which takes place in the mirage, and also in some respects similar to an action which wind at high altitudes exercises on waves of sound. Hence it may be that the electric radiation sent out from an antenna which is radiated in an upwards direction or in a more or less horizontal direction will be bent downwards so as to strike the earth at a region which falls short of the distant receiving station. This may be the explanation of the reduction of signalling distance by day as compared with that by night.

Later observations by Marconi have revealed the curious fact that the obstructive effect on transatlantic radiotelegraphy by this so-called "daylight effect" is greatest when the daylight or dark-·ness extends only part of the way across the ocean, one station being in day and the other still in night.

Another matter of great importance is the effect of obstacles, and especially mountains and earth curvature, upon long distance radiotelegraphic transmission. Even in the case of ordinary short distance work the effects produced by interposed hills and cliffs are quite marked.

A large number of observations have been recorded on this matter by Admiral Sir Henry Jackson, and his results were communicated to the Royal Society of London, in 1902. The experiments were conducted between ships of the British Navy provided with apparatus on the Marconi system, and the observations proved that the interposition of land, especially rocks of certain kind, greatly reduces the maximum signalling distance as compared with the distance for the same power for open sea.

Summarising the results for soft rocks, hard limestone, and limestone containing a large proportion of iron ores, respectively, the percentage of maximum signalling distance through them, compared with the open sea distance, is as follows :—

Sand, sandstone, shale, etc.		Hard limestone.	Iron ores.
Maximum distance	81	68	Less than　40
Minimum　　,,	56	25	,,　　　23
Mean　　　,,	72	58	,,　　　32

The results obtained show conclusively that hard rocks containing iron ores, interposed between the transmitting and receiving stations, especially when in the form of high cliffs, undoubtedly exercise a very marked effect in reducing the possible signalling distance for given types of ordinary short distance apparatus. Even when the cliffs do not extend to any height, there is evidence that the passage of the wave over land weakens its energy. This being, of course, merely a limited case of the more general fact that, for a given expenditure of energy, radiotelegraphy can be conducted over a greater distance over sea water than over dry land, the reasons for which have already been discussed.

Another familiar cause of disturbance and limitation of range is found in atmospheric electrical states or conditions, which often constitute a most serious obstacle to effective transmission, even over moderate distances. They are much less frequently noticed in temperate than in sub-tropical regions. In the Mediterranean Sea they seem to be particularly prevalent, and are most persistent in summer and autumn. Owing to their sudden advent and sudden cessation, it is difficult to carry out systematic or prolonged experiments. As already mentioned, these atmospheric disturbances exhibit themselves by making irregular automatic records upon the tape in the printing apparatus, or sounds in the telephone with a telephonic receiver. Admiral Jackson mentions that one of the most frequently recorded of these atmospheric markings is three dots, with a space between the first two, like the letters E I on the Morse Code, and this is very often due to distant lightning. Such disturbances are more frequent in summer and autumn than in winter and spring, and in the neighbourhood of high mountains than over the open sea, and with a falling barometer than with a rising one. In certain fine weather they reach their maximum between 8 and 10 p.m., and frequently last the whole night, with a minimum of disturbance between 1 a.m. and 1 p.m. When these atmospheric disturbances are present, the actual working distance of radiotelegraphy for any given apparatus may be reduced from 50 to 80 per cent. compared with that in perfectly clear weather. Thus, two ships, whose sea signalling distance may be 65 to 70 miles on a calm, bright day, may hardly be able to communicate 20 miles at a time when atmospheric disturbances are frequent. Admiral Jackson also notes the disturbing effect produced by a dry wind, such as the sirocco.

On the other hand, it is a matter of common experience that in certain conditions of the atmosphere the expenditure of an extremely small amount of power in the production of electric waves of the right length is effective in creating signals on a syntonic receiver at immense distances, and these occasional feats are not due to special skill on the part of the operator, but to favourable atmospheric conditions.

We have yet to mention another remarkable fact in connection with long distance radiotelegraphy, and that is the small degree to which the curvature of the earth seems to affect intercommunication between stations which employ earth-connected antennæ.

It is well known that rays of light and sound are diffracted to some extent round obstacles, but the long Hertzian waves from an earthed antenna appear to pass a quarter of the way round the earth without extravagant diminution of amplitude other than that due to distance and atmospheric absorption. We cannot compare experimentally the power required to send such waves 3000 miles over a flat surface with that required to send them 3000 miles round the earth; but the increase, be it what it may, is not such as to make terrestrial radiotelegraphy on a large scale impossible. It has been suggested that the conductivity of the upper layers of our atmosphere is sufficiently great to confine the waves to a spherical shell of the lower atmosphere, but the data at present available are not sufficient to entirely confirm this conclusion.

The hypothesis of ionic refraction above mentioned has been invoked also to explain this long distance radiotelegraphy. It seems now to be certain from mathematical investigations that true diffraction will not entirely explain the bending which takes place in the electric rays which are sent out from long distance radiotelegraphic stations. If, however, the electric wave travels faster in the upper levels of the atmospheric in consequence of ions present there, due to ultra violet light or to electrons projected from the sun, then such action is in the right direction to assist in bending the ray so as to follow round the curvature of the earth. It is only in some such way that we can account for the transmission of radiotelegraphic signals a quarter of the way round the globe, and also the increased night-time range. The latest views on this subject seem to be that diffraction alone may possibly account for practically all the effect at large distances, say 2000 or 3000 miles during the daytime; but that we have to call to our aid some such theory as that of ionic refraction to account for the greater irregularities and greater signal range which is found to exist by night.

The power of radiotelegraphic communication is greatly disturbed at times by stray or vagrant electric waves produced by atmospheric discharges. These are called Xs or strays, or atmospherics. When an automatic recording instrument is used such as a Morse inker or photographic recorder, these strays make false signals by printing irregular dots or dashes on the tape which confuse the message bearing signals. When the telephonic or aural method of reception is employed together with a high spark frequency 300 to 500 per second, then it is comparatively easy for the operator to separate out by ear the true signals which give a clear musical note from the false signals which give grunts, squeaks, or fizzles on the telephone. A great many devices have been invented for filtering out atmospherics. They are more obtrusive by night than by day, in summer than in winter, and in the tropics than in higher latitudes.

CHAPTER VIII

1. Radiotelegraphic Measurements.—As soon as radiotelegraphy emerged from its earlier stages, the importance of quantitative measurements in connection with it became evident. Any branch of knowledge only becomes science in proportion as it becomes the subject of exact measurements. Hence when operating with radiotelegraphic apparatus involving condensers, inductance coils, and other appliances, it is necessary to have the means of measuring and expressing precisely the magnitude of these quantities in terms of certain units, and to state our capacities, inductances, frequencies, wave lengths and decrements, in numerical values. The radiotelegraphist is therefore not equipped for his work unless he has a knowledge of the manner in which these measurements are made, and of the appliances necessary in making them.

We shall consider in turn the principal important measurements which have thus to be made.

2. The Measurement of High Frequency Currents.—In dealing with electrical oscillations we have to consider not merely the value of the current at any instant, but what is generally more important, its mean-square value, which is the only value capable of being directly measured. Supposing that a single train of oscillations is sent through a fine wire, it would expend part or the whole of its energy in heating the wire, and if a rapid succession of trains of oscillations were sent through the wire they would create heat in it at a certain rate, which would be balanced by the radiation of heat from the wire, or by the removal of heat in other ways, such as by the convection of the air. Since the emissivity of the wire increases with the temperature, a wire in any given surroundings subject to such oscillations will at last attain a steady temperature. Supposing, then, that we pass through the same wire a continuous current having a certain value, J, adjusted to produce the same temperature in the same wire under the same circumstances—that is, to produce the same quantity of heat as the successive trains of oscillations per second. The current J is called the root-mean-square value of the oscillations, and the square of J is called the mean-square or integral value of the oscillations. We can therefore find the root-mean-square value by passing the oscillations, whether damped or undamped, through a hot wire ammeter, provided that this has a suitable form. As already explained, electrical oscillations concentrate at the surface of the wire, and hence the true resistance of the wire to these oscillations is higher than its resistance to steady currents, by an amount depending on its section. If, however, we employ a wire not larger than No. 36 S.W.G., or a number of such wires in parallel, the high frequency resistance will practically be identical with the steady or ordinary resistance. Accordingly, an ammeter for high frequency currents or oscillations must be constructed in the following manner. It must consist of one or more bare fine wires of any material, which may be copper, platinoid, or constantan, and these wires should not be too tightly twisted together, but somewhat spaced apart. They must be contained in an enclosure such that when heat is produced in the wire at a certain rate it will, after a short time, come to a constant temperature. The root-mean-square value of any high frequency current is then measured by the value of the steady current which will bring the wire or wires to the same temperature. We may ascertain this temperature in one of three ways.

(1) We may make use of the expansion of the wire itself, so that when expanded to the same extent it is taken to be at the same temperature. As this expansion is always small, it is best measured by the sag produced when the wire is held between two fixed terminals. For this purpose, the wire or wires are attached to two insulated terminals, A and B, which should be mounted on marble or slate. To the centre of the bunch of wires is attached one end of another long wire which has its second terminal also fastened to a fixed point, C. From the centre of this last wire another fine wire or thread is attached to the end of an index needle (see Fig. 1). If then the first wire AB expands and the ends being fixed it sags up or down, this sag causes the second wire DC to sag in the middle still more, and that motion is multiplied by the index needle. In this way a very small increase in length of the wire AB is readily made evident by a movement of the needle. The instrument is then calibrated by passing through it certain measured steady or continuous currents, and noting the position on the scale at which the needle finally stands. The scale may thus be graduated in amperes or milliamperes. If then

we pass through the wire AB electric oscillations, either damped or undamped, the wire will be heated and the needle will take a certain position on the scale, and the scale reading gives the root-mean-square value of these oscillations.

(2) We may determine when the wire has the same temperature for oscillations or for steady currents by enclosing it in the bulb of an air thermometer (see Fig. 2). German investigators have made a good deal of use of this form of hot wire ammeter, which they generally describe under the name of a Reiss electrical thermometer. The instrument was in fact, however, invented by our countryman, Sir William Snow Harris, and described by him in 1827 in the *Philosophical Transactions of the Royal Society.*

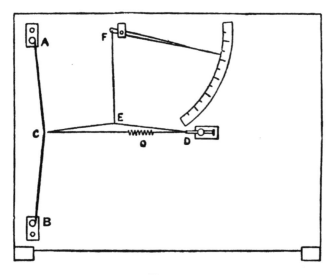

FIG. 1.

In this instrument, a fine wire or number of fine wires are included in the bulb of an air thermometer, consisting of a U-tube attached to a bulb, the bend of the tube being filled with some liquid. When a current is sent through the wire, it heats the wire and the air, and forces the liquid up one leg of the U-tube until a stationary position is reached. The instrument may therefore be graduated by passing various known continuous currents through the wire. To eliminate errors due to changes of external temperature, it is necessary to make the instrument in the form of a differential thermometer with two bulbs, in one of which the wire traversed by the currents is fixed.

(3) Another method of determining the temperature of the wires heated is by attaching to them a thermoelectric junction, and connecting the latter to any form of sensitive galvano- meter or voltmeter. If we employ a voltmeter or galvano- meter, the scale divisions of which are of equal length and indicate equal increments of current through the galvanometer, then, using a thermo-junction, say, of bismuth and iron pressed against the centre of a fine wire, it will be found on passing continuous currents through this wire that the scale readings of the galvanometer attached to the thermo-junction are very nearly proportional to the square of the current passing through the fine wire. They are almost exactly proportional to the square of the current, if the wire and thermo-junction are enclosed in a vacuum.

In any case, such an arrangement affords the means at once of determining the root-mean-square value of trains of oscillations sent through the fine wire, after the instrument has been calibrated by means of continuous currents of known strength.

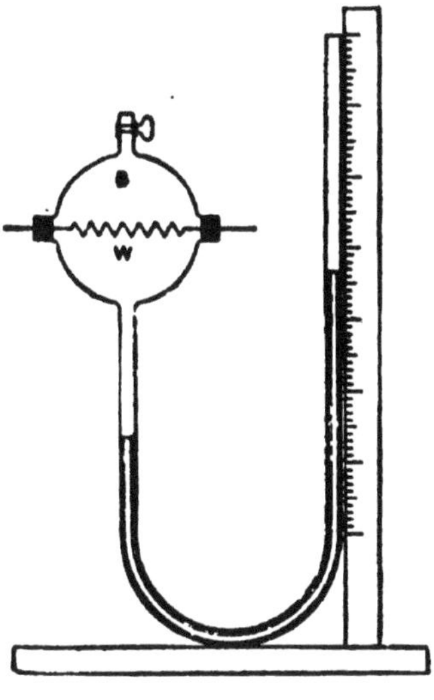

When we are operating with trains of damped oscillations which succeed each other at uniform intervals N per second, and if the oscillations them- selves have a frequency n and a decre- ment δ, then it can be shown that there is a relation between the root-mean- square value J of the oscillations and the first or maximum value I, in accord- ance with the equation

$$J = \sqrt{\frac{N}{8n\delta}} . I$$

Fig. 2.

Hence from the root-mean-square value and the known number of oscillations per train and their frequency we can determine the maximum value.

It is very astonishing to find the large values of the maximum currents that are reached during the trains of oscillations of an ordinary Leyden jar. Taking for instance a condenser having a capacity of $\frac{1}{400}$ microfarad, charged to a voltage of 12,000 volts, equivalent to a 3-millimetre spark, and discharged 50 times a second through a circuit having an inductance of 2000 cms., we find that the frequency of the oscillations is $2\,25 \times 10^6$, and the maximum current reached during the first half oscillation is as much as 420 amperes, whilst the root-mean-square value of the discharge currents would be only 1·5 amperes.

To measure large high frequency currents the Author has designed the following form of hot wire ammeter. It consists of a number of short fine wires, say of copper, not larger than No. 36, S.W.G. size. These are connected between two discs of brass or copper, so that the whole arrangement resembles a barrel, in which the fine wires are the staves. These discs are supported on brackets and connected to two terminals. The high frequency current then flows through all these wires in parallel, and the number of wires is adjusted to the currents to be measured. To one of these fine wires is attached a thermoelectric junction, say, of iron and constantan, and this junction is connected to a sensitive direct current ammeter. The whole arrangement is calibrated by continuous currents, by passing known direct currents from a battery through the fine wires, and noting the corresponding scale deflections on the thermoelectric ammeter.

The hot wire ammeter can then be applied to measure high frequency currents of any frequency within the limits of the reading of the ammeter connected to the thermoelectric junction.

The above-described hot wire ammeter can be constructed to measure currents of any value down to 1 or 2 milliamperes, but when smaller high frequency currents have to be measured we can employ the thermoammeter of Duddell which has been described in Chapter VI.

The disadvantage of all thermal ammeters is that since the heat created varies as the square of the current the scale indications of the instrument falls off very rapidly as the current decreases. It is, therefore, difficult to construct such instruments for extremely small currents.

In measuring the currents of the order of 40 or 50 microamperes which exist in the receiving antennæ of wireless plants we have therefore to adopt various devices. We may employ an Enithoven galvanometer with some form of rectifier. The above-mentioned instrument consists of a fine fibre of quartz or glass, which is silvered to make it conduct. It is stretched in the field of a very powerful electromagnet with the fibre at right angles to the lines of force. When a continuous or direct current passes through the fibre it is displaced across the field, and this displacement is observed with a microscope. If the fibre is placed in series with a crystal of carborundum or other rectifier the combination becomes able to detect very feeble alternating high frequency currents of a few microamperes. It can be calibrated by causing measured larger currents of the same frequency to affect the galvanometer through a step down air core transformer which reduces their value in some known ratio.

3. Measurement of Potential Difference.—In measuring the potential difference of points on a circuit traversed by oscillations, we may in the same manner desire to know either the maximum value reached by this potential difference at any moment or its root-mean-square value. The only method for obtaining the maximum value at any moment is by measuring the spark length of the spark which can be taken between balls of a known size connected to these two points respectively.

We have in Chapter II. given a table showing the spark voltages of sparks of various lengths between balls 2 cms. in diameter. Hence, if two such balls are connected to the points in question, and adjusted to such length that sparks will just not pass, we can obtain from the table an approximate estimate of the maximum voltage between these balls.

Nevertheless, the measurements can only be at the best approximate, because if many sparks are allowed to pass the ball surfaces become heated, which tends to promote discharge of a lower voltage, also the air becomes altered in constitution between the balls, acting in the same manner. Again, if daylight or the light from other sparks is allowed to fall upon the balls, the effect of the ultra-violet light contained in such luminous radiation also promotes discharge at a lower voltage. Nevertheless, the spark method is almost the only method we possess for determining the maximum of potential difference reached during a train, or at any time between two points on a circuit. The root-mean-square value of the voltage is best measured by means of an electrostatic voltmeter, the capacity of which is as small as possible. A voltmeter of the quadrant type, having a needle or movable plate suspended by a quartz fibre, is sometimes used for this purpose, such as that devised by Dolezalek.

The voltmeter method, however, must be used with caution, because the capacity of the voltmeter itself is sufficient to seriously disturb the condition of the circuit owing to the potential difference which exists when the voltmeter is removed.

Under some circumstances we can make use of such a voltmeter to measure the maximum voltage between two points, since there is a relation between the maximum value and root-mean-square value of the voltage, similar to that which exists in the case of the oscillatory current. In other words, if V is the maximum potential existing between two points of a circuit in which oscillations are taking place, having a frequency n, the trains succeeding each other at the rate of N per second, then if

U is the root-mean-square value of the voltage, U and V are related together by the formula

$$U = \sqrt{\frac{N}{8 n \delta}} \cdot V$$

4. The Measurement of Capacity.—One of the most frequently needed measurements in connection with this subject is the exact measurement of capacity. The unit of capacity in the electro-static system is the capacity of a conducting sphere 1 cm. in radius when placed in space at a considerable distance from all other conductors. Capacity is defined, as already mentioned, by the quantity of electricity necessary to charge the body to unit potential. Hence a sphere of 1 cm. radius is charged to a potential of one electrostatic unit (equal to 300 volts) by placing upon its surface one electrostatic unit of quantity, since all portions of the charge are then at a distance of 1 cm. from the centre, and create there a potential of one unit, and therefore raise the whole sphere to the same potential. The capacity of any other condenser can therefore be expressed in electrostatic units by stating its values in centimetres, meaning by that the radius in centimetres of a sphere, the capacity of which is equal to that of the condenser in question. The practical unit of capacity for most purposes is, however, the microfarad, or the millionth part of one farad. The microfarad is a capacity nearly equal to 900,000 electrostatic units. In other words, the microfarad is the capacity of a conducting sphere in free space, the radius of which is 9000 metres, or rather more than five and a half miles. The capacity of the whole earth in space, considered as a sphere, is only equal to 800 microfarads, and hence the unit of capacity called a farad, equal to a million microfarads, is far too large a unit for any capacities which have to be measured by terrestrial electricians. The farad is, however, the practical unit of capacity in consistent relation with the ampere, the ohm, the watt, and the joule, and therefore capacities in any other unit have to be reduced very often to capacities in farads, in substituting their numerical values in equations.

A convenient practical unit for radiotelegraphic purposes is the micromicrofarad, that is, a millionth part of a microfarad, equal to nine-tenths of an electrostatic unit. Capacities will therefore be sometimes measured in electrostatic units or micro-microfarads, and at other times in microfarads, depending on their magnitude.

The reader will find, in text-books on Physics, a large number of methods given for the comparison of capacities. We shall consider only those that are well adapted for radiotelegraphic purposes.

Two methods may be employed, one of which is a *comparison method*, which assumes the possession of another condenser of known capacity, and the other is an *absolute method*, in which the capacity is determined with reference to absolute units of resistance and time, without reference to any other condenser. The comparison methods are generally employed where rather large capacities, something of the order of a microfarad, have to be measured.

In every well equipped laboratory will be found a standard microfarad condenser, or half microfarad, generally constructed with mica as a dielectric interleaved between tinfoil sheets. Supposing, then, that we have another condenser, differing not much in capacity from one microfarad, or, at most, a moderate fraction or multiple of it, we may proceed in the following manner. If the standard condenser is charged by connecting its terminals with those of a voltaic cell or battery of cells, it takes up a certain quantity of electricity, Q, measured by the product of its capacity, C, and the voltage, V, to which it is charged. If, then, we connect this condenser to a galvanometer, a charge rushes out of the condenser through the galvanometer and causes the coil or needle to move through an angular deflection called the "throw." It can be shown that the sine of half the angle of the throw is proportional to the quantity of electricity that passes through the galvanometer, and for small deflections this may be taken as proportional to the displacement of the spot of light on the scale, if the galvanometer is a mirror galvanometer. If, then, two condensers of different capacity are charged to the same potential, and successively connected with the same galvanometer, the throws obtained will be nearly proportional to the capacities of these condensers. In order to eliminate error due to the want of precise proportionality of the throw to the quantity of electricity, we may proceed as follows :—

Connect the terminals of a voltaic cell or battery to a very high resistance, divided into sections, and connect the standard condenser across any fraction, R, of this resistance, and then connect the condenser immediately afterwards with the galvanometer, and observe the throw. Perform the same experiment with the other condenser, and vary the resistance between the terminals of the condenser, so as to find by trial the value it must have in order that the throws may be the same in the two cases. Then calling the two resistance values R and R', and the two capacities C and C', the capacities are inversely proportional to the value of the resistances.

The method is more conveniently carried out by arranging two capacities C_1, C_2 and two resistances, R_1, R_2, as in Fig. 3, with the battery and the galvanometer in opposite diagonals. A key is placed in the battery circuit, and the resistances, R_1, R_2, are altered until, on raising and lowering the key, the galvanometer gives no deflection. Under these circumstances, the capacities are inversely as the resistances, or

$$\frac{C_1}{C_2} = \frac{R_2}{R_1}$$

This method is known as De Sauty's method for the comparison of condensers.

· One source of difficulty in connection with it is found in the unequal absorptions of the dielectrics of the two condensers, if these are made of different materials. When a condenser is charged for a certain time and then discharged, a certain proportion of the charge comes out instantly ; the remainder comes out more slowly, and is called the residual charge, or the absorbed charge. Different dielectrics exhibit this effect in different degrees ; hence, if one of the condensers has a dielectric, say, of paraffin paper, and the other of glass or mica, it is sometimes difficult to find any ratio of the resistances which entirely abolishes all movement of the galvanometer needle when the battery key is raised or lowered.

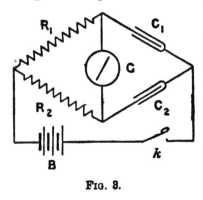

Fig. 8.

For a further discussion of these difficulties the reader may be referred to the author's "Handbook of the Electrical Laboratory and Testing Room," vol. 2, chap. ii.

The majority of capacities which we have to measure in radiotelegraphy are fractions of a microfarad, and for this purpose the most convenient method is an absolute method, in which the charge put into the condenser by a given voltage is repeatedly discharged through a galvanometer a known number of times per second, and measured as an electric current. If, for instance, one terminal of a battery, condenser and galvanometer are connected together and if the other terminal of the condenser is alternately connected to the terminal of the battery, and to that of the galvanometer, and if this process is repeated rapidly n times per

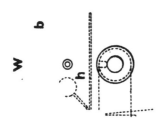

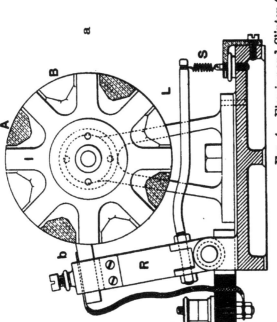

Fig. 4.—Fleming and Clinton Commutator for absolute capacity measurement.

second, then the galvanometer is traversed by a series of discharges, each of which conveys a quantity of electricity equal to CV microcoulombs, where C is the capacity of the condenser in microfarads, and V the potential of the battery in volts. If these n discharges succeed each other rapidly, the effect on the galvanometer is equal to the passage through it of a current equal to CVn microamperes, or $\dfrac{CVn}{10^6}$ amperes. This process of rapidly charging and discharging the condenser can be most conveniently effected by means of a rotating commutator, the structure of which is shown in Fig. 4. It consists of an electric motor of $\frac{1}{4}$-H.P., with associated starting and controlling resistances which can be run off in constant voltage circuit of 100 or 200 volts, with a constant speed which may conveniently be about 1500 R.P.M. To the shaft of this motor is connected by a flexible coupling the commutator, which

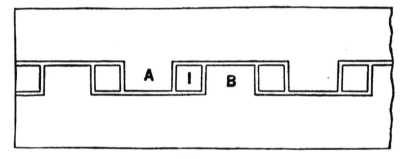

FIG. 5.

consists of two gun-metal discs, A, B, each having four projecting lugs like a crown wheel, and between these is placed another wheel, I. These three wheels are keyed together upon a shaft, and insulated from one another, and they form a drum, the surface of which appears as in Fig. 5 on looking down upon it. Against this drum three brass wire brushes press, which are carried on insulating pillars, R. Two of these make a contact with the outer flanges of the crown wheels, and the middle one makes contact with the central portion. It will be seen, therefore, that as the drum revolves, the centre brush alternately makes contact four times in each revolution, or a hundred times per second, first with the brush on the right hand, and then with the brush on the left hand. The time of contact is also accurately known, and there is no bouncing or uncertain contact, but a smooth, steady contact. The speed of the commutator is determined by attaching to the shaft an endless screw intergeared with a toothed wheel, so that the wheel makes one revolution for every hundred revolutions of the shaft. At each revolution of the wheel a pin lifts a·lever, which strikes a blow on a gong. By means of a stop watch the time of ten revolutions of the wheel, and therefore a thousand revolutions of the commutator, can be ascertained with an accuracy of less than one per cent. If, then, we connect one terminal of a condenser to the middle brush, whilst the two outside brushes are connected respectively to the terminals of a galvanometer and of a commutator, the other terminals of the galvanometer, commutator, and condenser being connected together, when the commutator is set rotating, it will cause a series of discharges to take place in the galvanometer, which will have all the effect of a steady current and create a deflection which remains constant. To determine the value of the steady current, which will give the same deflection, we may proceed as follows :

Place a large resistance, R, in series with a galvanometer and a small shunt, S, across the terminals, and let G be the resistance of the galvanometer itself. Apply the same battery used in charging the condenser to the terminals of the galvanometer, and alter the value of the large resistance R to the shunt S until the galvanometer gives the same deflection with a steady current passing through it as it did with the intermittent series of condenser discharges. Under these circumstances we have the following equation between the capacities, resistances, and number of discharges, viz :

$$\frac{n\text{VC}}{10^6} = \frac{\text{V}}{\text{R} + \dfrac{\text{GS}}{\text{G} + \text{S}}} \cdot \frac{\text{S}}{\text{G} + \text{S}}$$

from which we deduce

$$\text{C} = \frac{\text{S} \cdot 10^6}{n\text{R}(\text{G} + \text{S}) + n\text{GS}}$$

The value of n, or the number of discharges per second, is accurately determined by counting the number of revolutions per second of the commutator, and multiplying them by four. The voltage of the battery used in this experiment must be determined by the magnitude of the capacity to be measured. If that capacity is a very small one, then it may be necessary to use a well insulated battery of a hundred small secondary cells, whereas, if the capacity is very large, one cell may suffice.

The above method is very well adapted for determining the capacity of an antenna. In this case the antenna is connected to the middle brush, the two outer brushes pressing against the commutator shown in Fig. 4 being connected respectively to the galvanometer and to one terminal of a well insulated battery of a hundred small secondary cells. The other terminals of the galvanometer and battery must be connected to the earth. On rotating the commutator, the antenna is alternately charged by the battery and discharged through the galvanometer, and this capacity may be determined in microfarads as above described.

In using this commutator for the measurement of very small capacities, such as an antenna, it is necessary to take into account the capacity of the commutator itself, which is not altogether negligible. This is done by taking two readings, one with the antenna connected to the middle brush, and the other with the antenna removed, and the difference of the capacities determined in the two cases may be taken to be that of the antenna. By this means we can easily measure the capacity of an antenna consisting of a wire 0·1 inch in diameter and 100 feet long elevated into the air, a capacity which will generally be found to be approximately $\frac{1}{5000}$ of a microfarad.

The above method is also a convenient one for measuring the capacity of Leyden jars, or similar condensers, which have a capacity of the order of 0·01 to 0·002 or thereabouts of a microfarad. In the case of Leyden jars it should, however, be noted that the actual capacity when used with high frequency potentials is increased by the effect of the glow discharge from the edges of the tinfoil.

5. **The Measurement of Inductance.**—Another important measurement is that of inductance, which can be measured either absolutely or by comparison with certain standards of inductance. Innumerable methods have been described for the measurement of inductance; but amongst those with which the author is acquainted, none is simpler or more easily carried out than that due to Professor A. Anderson.

Anderson's method is applicable in the determination of inductances from a few millihenrys up to any multiple of a henry, and by the adoption of certain modifications suggested by the author is capable of measuring inductances as small as a few microhenrys. In its simplest form it is carried out as follows :—

The coil of which the inductance is desired, marked RL in the diagram in Fig 6, is connected with three other resistances, so as to form a Wheatstone's bridge arrangement. In the circuit of the battery B is placed a key k, and in the circuit of the galvanometer is placed another key, k', and the galvanometer is joined in series with a resistance, *r*, and also a condenser, C, is connected between one terminal of the galvanometer and one angle of the bridge.   Let R be the resistance of the coil of which the inductance is required reckoned in ohms, and L its inductance in henrys, and let PQS be the resistance of the other arms of the bridge.   Then the first step in the process is to vary the values of P, Q and S, so that when the battery key is first put down, and afterwards the galvanometer key, there is no movement of the galvanometer coil or needle.   The bridge is then said to be balanced by steady currents.   The resistance S should be a plug-box resistance running from 0 to 10,000 or 20,000 ohms, and the

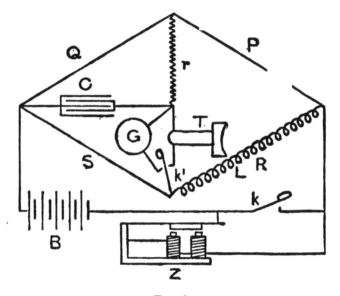

FIG. 6.

condenser C should preferably be a variable capacity.   In obtaining a steady balance of the bridge, the resistance *r* should be cut out, and the condenser C removed.   When the balance is obtained these instruments are re-inserted, and we then find that if the galvanometer key is first closed, and afterwards the battery key, the galvanometer coil or needle gives a throw or movement, which, however, can be entirely annulled by suitably varying the capacity C or the resistance *r*.   When this is the case, the bridge is said to be balanced for throws.   The advantage of the Anderson method is that in obtaining the balance for throws we do not have to upset the steady balance previously obtained.   When the steady balance is obtained, there is a proportionality between the four resistances P, Q, R, S, expressed by the formula:

$$\frac{P}{Q} = \frac{R}{S}$$

and when the balance for throws is obtained, there is a relation between these resistances and the inductance L and capacity C and resistance r, as follows:

$$L = C\{r(R + S) + QR\}$$

For the proof of this formula the student is referred to the author's "Electrical Laboratory Notes and Forms," Form No. 47, or to the "Handbook for the Electrical Laboratory Testing Room," vol. 2, chap. ii., p. 193.

The student should note that in the above formula, if P, Q, R, S and r are measured in ohms, and C in microfarads, then the inductance L will be given in the above formula in microhenrys, and must be divided by 1000 to reduce it to millihenrys, or by 1,000,000 to reduce it to henrys. It will be found that when the inductance is very small, the throw of the galvanometer is small also, and there will be considerable difficulty in determining when it vanishes. We may, however, increase the sensibility of the method considerably by the following plan, due to the author: In the circuit of the battery is placed a vibrating electro-magnet or buzzer, Z (see Fig. 6), which continually interrupts the battery current, say two or three hundred times a second. We then insert in parallel with the galvanometer a telephone and a throw-over switch. During the operation of obtaining the steady balance, the buzzer is cut out and the galvanometer introduced into the bridge circuit. When this is done the telephone is substituted for the galvanometer, and the buzzer is re-inserted, and the observer then alters the resistance of r, or the capacity of the condenser, until no sound is heard in the telephone, or, at least, a minimum sound. To do this exactly, the buzzer must be enclosed in a sound-proof box placed in a cupboard, or at a distance from the observer, or otherwise he will mistake the sound due to the buzzer for a sound in the telephone. Those who have acute hearing can carry out the measurement with great accuracy, and in this way measure the inductance down to a few microhenrys. To check the accuracy of this measurement, it is desirable to possess certain standards of inductance of known value. These are best constructed by forming a square circuit of one single turn of round copper wire, say No. 16 S.W.G. Two long wooden laths are arranged in the form of a cross, and wires strained round them so as to make a square circuit interrupted at the ends of one diagonal. From the formula

$$L = \frac{8S}{1000}\left(2{\cdot}3026 \log_{10} \frac{16S}{d} - 2{\cdot}853\right)$$

already given in Chapter I., we can predetermine the inductance of this circuit for high frequency currents, knowing the length S of the side of the square, and the diameter d of the wire, both measured in centimetres.

The above formula gives us the inductance in microhenrys, each of which is equal to one thousand absolute C.G.S. units of inductance, which are reckoned in centimetres. Where great accuracy is not required, a larger standard of inductance can be constructed by employing the formula due to Russell for the inductance of a spiral circuit. If a bare No. 16 S.W.G. copper wire is wound upon a round rod of ebonite or hard wood in a helical form, either by cutting a screw groove in the wood or else winding a silk string in between the turns of the wire, and if the mean diameter of one turn of the helix is D centimetres, and l is the length of the spiral in centimetres, and N the number of turns per centimetre, then the inductance of the coil in microhenrys is given by the formula

$$L = \frac{(\pi DN)^2 l}{1000}\left\{1 - 0{\cdot}424\left(\frac{D}{l}\right) + 0{\cdot}125\left(\frac{D}{l}\right)^2 - 0{\cdot}0156\left(\frac{D}{l}\right)^4\right\}$$

The formula, however, neglects to take account of the fact that in the case of spiral wires traversed by high frequency electric oscillations the tendency of the current to concentrate on the inner sides of the coils tends to increase the inductance by a small amount.

Anderson's method is also applicable for the measurement of mutual inductances. If two coils are placed with their axes in one line, they exert on each other a mutual inductance, and a current in one produces an induced current in the other. The mutual inductance or coefficient of a mutual inductance, M, is defined to be the numerical value of the total magnetic flux which is linked with both coils when a unit steady electric current flows in them. Hence, if we have two coils with their planes parallel to one another, and we pass through them a steady unidirectional current, each coil is self-linked with a certain magnetic flux due to the current in it, and it is also linked with the magnetic flux due to the current in the other coil. The inductance of the coil or its coefficient of self-inductance may be defined to be the flux which is self-linked with its own circuit when unit current flows through that coil.　Hence, if we have two coils whose inductances are respectively L and N, and their mutual inductance M, and if they are joined in series, it is obvious that they may be so joined up that the current flows the same way round in both coils, or in the opposite direction of the two coils.　In the first case the circuit consisting of two coils is self-linked with a total flux proportional to L + 2M + N, and in the second case a flux proportional to L − 2M + N.　Accordingly, if we join up the two coils in the first manner, and measure by the Anderson method the inductance of the entire circuit and call it L_1, and then join it up in the second manner, and measure its inductance again and call it L_2, the difference of these inductances must be equal to four times the mutual inductance, or

$$M = \frac{L_1 - L_2}{4}$$

Hence, given two coils, we can measure separately their inductances, and also their mutual inductance in any position.　The quotient of the mutual inductance by the square root of the product of the separate inductances of the two coils, that is, the quantity $\frac{M}{\sqrt{LN}}$ is an important quantity called the *coefficient of coupling*, and is denoted by the symbol k.

The above methods, therefore, enable us to determine the coefficient of the coupling to the two coils which form the primary and secondary coils of a transformer of any kind.

6. **Measurement of Frequency.**—A third essential measurement is the measurement of the frequency of the oscillations, whether damped or undamped, in an oscillating circuit comprising a condenser or capacity and an inductance.

It has already been shown that the frequency of the oscillations in a circuit having a capacity of C microfarads and the inductance L microhenrys, is given by the formula :

$$n = \frac{10^6}{2\pi\sqrt{C_{\text{mfds}}\, L_{\text{mhys}}}}$$

or if the inductance is measured in absolute electromagnetic units or centimetres, then the frequency is given to the formula :

$$n = \frac{5{\cdot}033 \times 10^6}{\sqrt{C_{\text{mfds}}\, L_{\text{cms}}}}$$

If, then, we have a circuit possessing capacity and inductance in which oscillations are taking place, we can determine the frequency of these oscillations by the principle of resonance.

If we place near to but not very close to the circuit in question another circuit containing a known capacity and a known inductance which can be varied, and also have some means for determining when the oscillations induced in this second circuit are at their maximum value, we may cause the oscillations in the first circuit to induce others in the secondary or detecting circuit, and we can vary either the capacity or inductance, or both together, of this last circuit until the current in it has its maximum value. In this case the two circuits are said to be in resonance. It has already been shown that if the two circuits are closely coupled, oscillations of two frequencies are set up in these circuits, but if they are loosely coupled, the resonance curve is a curve with a single peak, and the current in the secondary or detecting circuit will have its maximum value when the capacity and inductance are so adjusted that the product of these two quantities for the secondary circuit is the same as the product of the two quantities for the first circuit, and then, knowing the capacity and inductance in the secondary or detecting circuit, we can determine the frequency n of the oscillations in the first circuit from the formula:

$$n = \frac{5 \cdot 033 \times 10^6}{\sqrt{C_{mfds} L_{cms}}} = \frac{5 \cdot 033 \times 10^6}{O}$$

The most convenient method of making these measurements is by means of an instrument devised by the author, called a Cymometer. This consists of a condenser of variable capacity constructed of a tube of brass covered with ebonite, on the outside of which another concentric tube fits closely, but not too tightly as to prevent easy movement. If the tubes lie over one another, such a double brass tube with interposed tube of ebonite constitutes a tubular condenser, but if the outer tube is slid off the inner brass tube the capacity is reduced almost proportionately to the displacement of the outer tube. Again, if we have a wire wound in the form of a helix round an ebonite tube, the turns being close together but not touching, and if we have some form of clip which can be slid along the helix so as to make use of more or less of the spiral, we have a variable inductance.

These two appliances are combined together in the cymometer in such a way as to form a complete oscillatory circuit; the inner end of the tubular condenser (see Fig. 7) is connected to one end of the helix of wire by a copper bar, and the outer condenser tube is connected to the helix by an embracing clip, so that as the outer condenser tube is displaced from the inner tube to reduce the capacity, the effective inductance in the circuit due to the spiral is reduced in the same proportion.   The helix and the tubular condenser, which may be formed of two or more tubes, are mounted on a board, and by means of a handle the condenser tube can be moved and the inductance and capacity simultaneously altered, and in the same proportion.   If, then, we place the long copper bar connecting the helix and condenser near but not very close to any other circuit in which oscillations are taking place, we can tune the cymometer circuit to the other circuit by moving the handle so as to vary the inductance and capacity of the cymometer.   We must then have some means of determining when the current in the

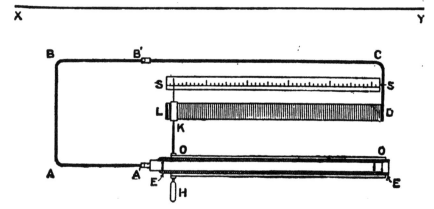

Fig. 7.

cymometer, or the potential difference of the tubes forming the condenser is a maximum.

· The author discovered that the most convenient way of doing this was by the use of a vacuum tube of the spectrum type, filled with Neon.   Neon is a rare gas contained in the atmosphere, about 80,000th part by volume, and it is remarkable for its small dielectric strength and for the great brilliancy of the glow produced in it when placed in an alternating current field.   If such a Neon tube is connected to the terminals of the tubular condenser, then when the capacity and inductance are altered and the oscillation in the cymometer circuit thereby increased up to a maximum, it is easy to determine the moment when this maximum takes place by the Neon tube beginning to glow, or glowing most brilliantly (see Fig. 8).

Another method of discovering when the current is a maximum in the cymometer circuit is by inserting in the circuit of the copper connecting bar a fine wire of high resistance about a centimetre in length, having in contact with it a very sensitive thermo-junction of bismuth and iron. This thermojunction is connected to a sensitive galvanometer, preferably a Paul single pivot low resistance galvanometer (see Fig. 9). If then by the movement of the handle of the cymometer it is gradually tuned with any adjacent circuit in which oscillations are taking place, the increase in the current up to a maximum will be indicated by a gradually increasing deflection of the galvanometer, and it is quite easy to determine that adjustment of the cymometer in which the current is a maximum.

The cymometer has a graduated scale with a pointer moving over it, and the instrument is calibrated by the manufacturer so as to show at a glance the frequency corresponding to any particular adjustment of the tubular condenser. The author has designed such instruments for reading frequencies from 50,000 up to 5,000,000, and the appearance of the complete instrument is as shown in Fig. 10.

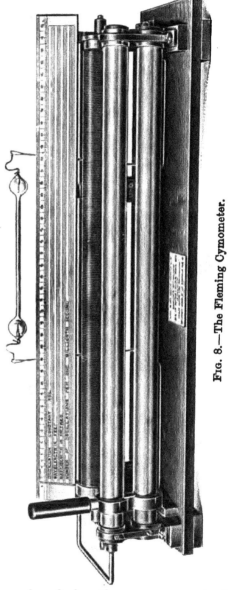

FIG. 8.—The Fleming Cymometer.

The cymometer may be employed for the measurement of small capacities and inductances in the following manner:—

Each instrument is, or can be, supplied with a standard

inductance consisting of one or more turns of insulated wire arranged round a rectangular frame. These inductances vary from about 4000 centimetres, or four microhenrys, up to 75,000 centimetres or 75 microhenrys, depending on the pattern of cymometer, in use. If then a certain small capacity, say, that of a Leyden jar, has to be determined, it is done in the following

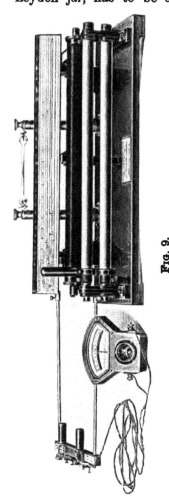

FIG. 9.

manner. The jar is placed upon a sheet of ebonite, and one coating is connected to one secondary spark ball of an induction coil, the other coating or terminal of the condenser being connected to one end of the above-mentioned standard inductance, whilst a second end of the standard inductance is connected to the other secondary spark ball (see Fig. 11). The spark gap, condenser, and inductance are all connected in series. The cymometer is then placed with its copper bar parallel, not very near to one side of the standard inductance. On working the coil, oscillations are set up in the circuit of the jar and inductance, and the handle of the cymometer is moved until the Neon tube glows most brightly. The scale reading of the cymometer then shows the oscillation constant of the cymometer in that position, that is to say, the value of the square root of the product of its capacity in microfarads, and its inductance in centimetres in its then position. The value of this quantity is called the *oscillation constant*, and is marked on the scale. It then follows that the oscillation constant for the circuit containing the unknown capacity must be the same.

Hence, if we square the value of the oscillation constant and divide by the value of the standard inductance in centimetres, we have the value of the unknown capacity in microfarads. Thus, for example, suppose that the standard inductance is 5000 centimetres, and that the maximum glow in the Neon tube occurs when the cymometer pointer indicates that the oscillation constant

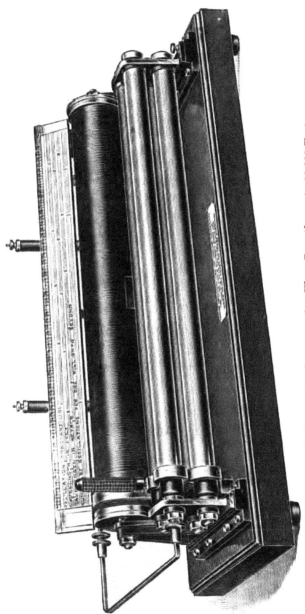

FIG. 10.—A Cymometer for measuring Wave Lengths up to 10,000 Feet.

is 10, then the square of 10 being 100, and the quotient of 100 ÷ 5000 being $\frac{1}{50}$, we know that the capacity of the condenser in question must be $\frac{1}{50}$ of a microfarad. The rule therefore is as follows : Square the oscillation constant and divide by the value of the standard inductance in centimetres, and the resulting quotient is the capacity of the jar or condenser in fractions of a microfarad.

In the same way the cymometer can be used with a standard condenser to determine the value of an unknown inductance, for if we determine as above described the capacity of a condenser by the aid of the cymometer, then join up this capacity with the unknown inductance and the spark gap, to form an oscillation circuit, putting in, if necessary, a yard of straight wire to lie parallel with the bar of the cymometer, and if we then determine

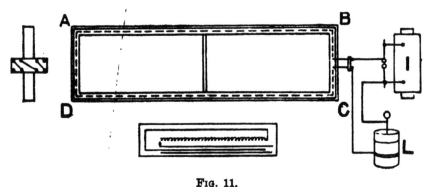

FIG. 11.

the oscillation constant of this circuit, and find it to be O, then the inductance in the circuit must be equal to $\frac{O^2}{C}$, where C is the capacity of the condenser in microfarads, and this quotient gives the inductance in centimetres.

In those cases where a small inductance is measured, it can be determined as the difference between two inductances, viz. by joining up with the condenser of known capacity a standard inductance of known value, and dividing the oscillation constant as above, and then increasing the inductance of that oscillation circuit by adding in the small unknown inductance, and making a redetermination of the oscillation constant. Supposing, for instance, that the oscillation constant in the first instance is O_1, and in the second O_2, and that the standard inductance was, say, 5000 cms., and the value of the unknown and small inductance L, then we have the following equations :

$$\frac{O_1{}^2}{C} = 5000$$

$$\frac{O_2{}^2}{C} = 5000 + L$$

$$L = \frac{O_2{}^2 - O_1{}^2}{C}$$

from which we can at once determine the value of L.

A large variety of such tests can be made with a cymometer, provided it is remembered that the oscillation constant marked on the scale of the cymometer is the square root of the product of its capacity reckoned in microfarads and its inductance in centimetres, corresponding to the position in which the handle of the cymometer is then placed.

7. **Measurement of Wave Lengths.** — Another important measurement is the measurement of the wave length of the waves emitted by an antenna, or of the wave lengths being received by an antenna. In all cases of wave motion there is a relation between the velocity of the wave V, its frequency n, and wave length λ, expressed by the equation

$$V = n\lambda$$

The velocity of the electromagnetic waves being 300 million metres per second, or very nearly one thousand million feet per second, it follows that the wave length is at once obtained by dividing this last number by the frequency. Hence, if the frequency of the oscillations in an antenna is determined, we have the wave length of the emitted waves. If, then, we can determine the oscillation constant of the antenna, or of the circuit which is radiating, we have at once the following rules :—

Wave length in feet = 195·56 × oscillation constant.

Wave length in metres = 59·6 × oscillation constant.

Frequency in millionths of a second is 5·033 ÷ oscillation constant.

In order to determine the wave length, therefore, all that is necessary is to place the bar of the cymometer parallel, but not very near to a portion of the lower part of the antenna. For this purpose, a yard or two of the antenna may be laid in a horizontal position, if necessary. On exciting the oscillations in the antenna and moving the handle of the cymometer, we shall find a position in which the Neon tube glows most brightly, provided the cymometer used has a range including the wave length in question. In the case of inductively coupled antennæ, it will be found, of course, that there are two wave lengths being emitted, and therefore two positions in which the Neon tube has a maximum glow. In so using the cymometer, it is desirable to put the bar as far as possible from the antenna after having roughly discovered the approximate wave length, and then to take a fresh reading, so adjusting the distance of the cymometer bar from the antenna, that the Neon tube only just glows on passing through to a position of resonance. With a little care it is possible to determine the wave lengths of the order of 1000 or 1500 feet within 10 feet.

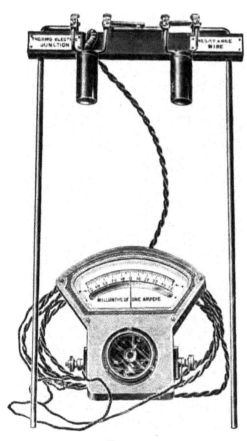

Fig. 12.

Four types of cymometers are now made, one suitable for measuring wave lengths from about 30 metres to 1000 metres, another up to 1500 metres, a third up to 2000 metres, and a fourth up to 3000 metres, the lowest possible reading being generally about one twelfth part of the highest possible reading for any one instrument, but by special means, greater ranges can be obtained. Hence a suitable cymometer must be employed for the particular measurements being made, the oscillation constants of the above four types ranging from about 1 to 12, 2 to 25, 3 to 37, and 4 to 50. For certain measurements in which greater accuracy of reading is required, it is better to employ, instead of the Neon tube, a thermoelectric detector, which is placed in the circuit of the cymometer. The circuit of the cymometer is cut in two places, or the simple double copper bend with which it is usually provided for completing the circuit can be replaced by a special double bend (see Fig. 12) containing two cuts in it, in one of which is inserted a fine resistance wire, and in the other a fine resistance wire having a thermoelectric junction in contact with it. These resistances and thermoelectric junction are contained in two ebonite boxes attached to the special bend, and a length of flexible connecting wire is provided, by which the thermoelectric junction is connected to a special low resistance single pivot sensitive galvanometer, that usually employed being made by Paul. There are short circuiting straps for cutting out the thermo-electric junction resistance, or the plain resistance. If we insert in the circuit only the resistance with the thermo-junction, and then employ the cymometer as above described, in proximity to any circuit in which oscillations are taking place, we shall find that as the handle is moved, tuning the cymometer more and more in circuit with the circuit under test, the ammeter exhibits a gradually increasing deflection, and at a certain position of the cymometer a maximum deflection is reached. In this position, therefore, the cymometer circuit is traversed by the maximum current, and, therefore, is in resonance with the circuit under test. In another form of wave-meter or cymometer devised by Dönitz, the condenser consists of a number of fixed plates interspaced between a number of movable plates attached to a shaft, by the rotation of which the plates can be more or less sandwiched in between each other, and the capacity of the condenser formed by these plates therefore varied within certain limits (see Fig. 13). This condenser is connected in series with certain coils of wire having a known inductance, and in addition a hot wire ammeter or electric thermometer, consisting of a wire enclosed in one bulb of an air thermometer, is employed as an indicating instrument to show when the current in the cymometer circuit is a maximum. The instrument has a scale to show the wave lengths or frequencies corresponding to any possible position of the rotating axis of the condenser; that is to say, of the capacity included in the wave-meter circuit. There is, however, a great advantage in employing an instrument like the author's cymometer, in which the capacity and the inductance of the instrument are varied simultaneously and in the same proportion, as then the divisions on the scale indicating the oscillation constant in various positions are equally spaced. It is also necessary to be able to measure the length of the waves which are incident on a receiving antenna. This may be done as follows :—

A single turn of wire is included in the receiving antenna

which is in inductive coupling with a standard inductance of known value. This inductance is in circuit with an oscillation detector, say, of the electrolytic type, and is shunted by a condenser of variable capacity. The capacity of the condenser is then varied until the coupled circuit is in resonance, and the standard inductance gradually moved away from the antenna coil, the condenser capacity being also varied to keep the tuning right. It will be found possible to put the standard inductance so far from the antenna coil that the slightest variation of the condenser

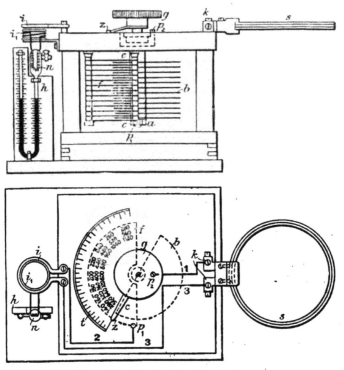

FIG. 18.—The Dönitz Wave Meter.

capacity either way causes the sound in the telephone to disappear. The capacity in microfarads is then noted, and also the value of the standard inductance in centimetres, and the wave length in metres obtained from the formula

$$\lambda = 59\ 6\sqrt{C_{\text{mfds}}\ L_{\text{cms}}}$$

8. Measurement of Damping and Logarithmic Decrements.— In connection with the production of electric oscillations by the spark method, a frequently needed measurement is that of the *decrement* of the oscillations. When damped oscillations exist in a circuit, they decay in amplitude according to the law that the ratio of any oscillation to the next preceding it is constant, and this constant ratio is called the damping of the oscillations, and the Napierian logarithm of the ratio of one oscillation to the preceding one, is called the logarithmic decrement, or shortly, the decrement. If we assume, as we may do, that the oscillations in a train are practically exhausted when the last oscillation is not more than one per cent. of the initial one, then, as already shown in Chapter I., the number of complete oscillations, M, in a train is given by the rule

$$M = \frac{4 \cdot 605 + \delta}{\delta}$$

The quantity δ is the logarithm of the ratio of two successive oscillations in the same directions to one another, or $2 \cdot 303$ times the ordinary logarithm to the base 10 of the same ratio. As far as regards a mere qualitative determination of the damping, that is a proof that the oscillations with which we are dealing are damped or undamped, probably the best method of doing it is by the vacuum tube oscillograph. This consists of a glass tube having two straight aluminium wire electrodes, in line with each other and nearly touching. The tube is exhausted, but only to a low vacuum, about equal to 1 mm. of mercury. When the electrodes of the tube are connected to the terminals of a condenser which is in electrical oscillation a glow appears on the electrodes, the length of that glow being proportional to the potential difference. If the tube is viewed in a rapidly rotating mirror, making, say, 500 revolutions a second, or at least a very high number, the alternately glowing electrodes produce separated images, and we can see at once whether the discharge is damped or undamped. The photographs in Figs. 14 and 15 are oscillographs, taken in this manner by Dr. Dieselhorst, of damped and undamped oscillations. In each of these the upper part is the positive and lower negative, and it will be seen that in the undamped oscillations, which are produced by a Poulsen arc, the oscillations have a greater amplitude on one side than on the other. This is because there is on the condenser a steady potential difference, which is that creating the current through the arc, and the oscillatory potential difference is superimposed on this, hence creating a non-symmetrical oscillogram.

As the quality of a train of electric waves and its effect upon a receiver greatly depends upon the damping, the determination of the quantity δ is an important measurement. It is easily effected by means of the cymometer, as follows:—

The cymometer circuit is, as explained, cut in two places, or else with an extra double bend of copper having two gaps in it, which takes the place of the ordinary simple double bend employed when using the instrument merely with the Neon tube. In these gaps can be inserted the two ebonite boxes which contain fine resistance wires of constantan, against one of which is pressed a fine bismuth or iron thermo-junction. To complete the output, we require a single pivot Paul galvanometer, having a resistance of about 4 or 5 ohms, and reading from zero up to 400 microamperes. Also it is requisite to have means of calibrating this instrument. Assuming the possession of a resistance box, a secondary cell, and a small direct-reading milliamperemeter, the

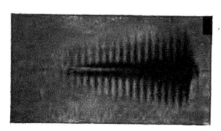
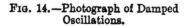
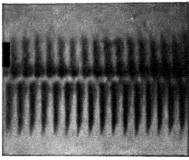

FIG. 14.—Photograph of Damped Oscillations.

FIG. 15.—Photograph of Undamped Oscillations.

first step is to calibrate the thermoelectric junction, so as to ascertain from the readings of the Paul galvanometer connected to the terminals of the thermo-junction the mean-square value of the current passing through the fine wire. For this purpose we connect the fine constantan wire in series with the cell, the milliamperemeter, and the variable resistance, and pass various currents through it, say from 1 to 100 milliamperes. The deflection of the Paul galvanometer connected to the thermo-junction is noted at the same time. Squaring the values of the continuous currents sent through the fine wire, we then plot a curve, the abscissæ in which represent the scale readings of the Paul galvanometer, and the ordinates the square of the value of the current through the fine wire producing this deflection. Hence, if we subsequently pass oscillations through the fine wire, the reading of the Paul galvanometer enables us to determine at once the mean-square value of these oscillations. It will generally be found that the curve connecting the squares of the currents passing through the hot wire with the deflection of the Paul galvanometer is practically a straight line. When this calibration is completed the fine wire with a thermo-junction in contact with it is placed in the circuit of the cymometer, and the bar of the cymometer is placed near to the circuit in which oscillations exist, the damping of which is required. In so doing, it is necessary to be careful not to bring the cymometer too near to the oscillation circuit under test at first, or else the oscillations set up in it may be so strong as to burn out the fine resistance wire in the ebonite box.

We proceed then to take a series of observations, as follows:— Set the cymometer handle at one end of the scale, so as to include all the capacity, and move it forward step by step, noting the reading of the oscillation constant for each stage, and at the same time the reading of the Paul galvanometer in connection with the thermo-junction. It will be found that on approaching the position of resonance the galvanometer reading will increase very rapidly to a maximum, and it may be necessary to make two or three rough trials, first adjusting the distance of the cymometer from the circuit under test until this maximum current in the cymometer makes a deflection of the galvanometer just within the range of the scale of the latter. We can then make a more careful experiment, plotting out a curve, the abscissæ of which are the oscillation constants, as read on the cymometer for each position of the handle, or the frequency n corresponding thereto, and the ordinates are the mean-square values of the currents in the cymometer circuit, as obtained from the readings of the Paul galvanometer and its curve of calibration. The curve so obtained is called a resonance curve, and it will be found to run up into a single peak very rapidly, unless oscillations of two frequencies occur in the circuit under test, in which case there will be two peaks (see Fig. 16).

Let a^2 be the mean-square value of the current in the cymometer, in any position of the handle corresponding to which the natural period of frequency of the cymometer is n_2, and let A^2 denote the mean-square value of the maximum current in the cymometer when it is tuned to resonance with the circuit under test, and let n_1 denote the corresponding frequency as read on the cymometer. Then the oscillation circuit under test has a certain decrement δ_1, and the cymometer itself has a certain decrement δ_2.

It has been shown by V. Bjerknes and P. Drude that the

following relation holds good between the decrements of the two circuits and the frequencies n_1 and n_2, viz. :

$$\delta_1 + \delta_2 = \pi \left(1 - \frac{n_2}{n_1} \right) \sqrt{\frac{a^2}{A^2 - a^2}}$$

provided that n_1 and n_2 do not differ from one another by more than, say, five per cent.

Since the frequency is connected with the oscillation constant

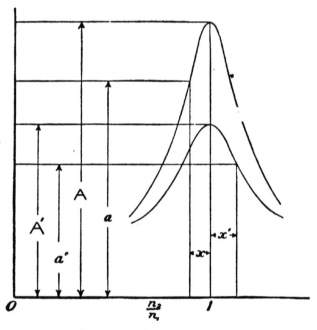

FIG. 16.—A Resonance Curve.

by the formula $n = \dfrac{5\cdot033 \times 10^6}{O}$, we can also write the above formula of Drude and Bjerknes in the following form :

$$\delta_1 + \delta_2 = 3\cdot1416 \frac{O_2 - O_1}{O_2} \sqrt{\frac{a^2}{A^2 - a^2}}$$

Plotting out the resonance curve as above described, it is best to take the mean-square value of the maximum curve as unity, and to correct the other currents in the corresponding ratio; and the same way for the frequencies, viz. the resonance frequency and any other frequency. If, then, we put x for $\left(1 - \dfrac{n_2}{n^1}\right)$ and y for $\dfrac{a}{A}$, we can write the above formula for the sum of the decrements finally in the form

$$\delta_1 + \delta_2 = 3\cdot1416\ x\ \frac{y}{\sqrt{1 - y^2}}$$

Since the resonance curve is not quite symmetrical with respect to its maximum ordinate, it is best to determine from the resonance curve the values of the frequency n_2 lying on either side of the maximum current, which correspond to any given value of the cymometer current, and to take the mean of these values as the value to be put into the above formula.

It will be seen, then, that from such a resonance curve we can determine the sum of the decrements of the circuit under test, and that of the cymometer. This last has, however, been increased by the resistance of the fine wire inserted in its circuit, by means of which we determine the sum of the decrements. We have therefore to eliminate the latter quantity as follows : It has been shown by Bjerknes and Drude that, if a secondary circuit has extra resistance of known value inserted in it so as to increase its decrement by a known amount δ_2', that the maximum current A (R.M.S. value) in the secondary circuit is altered to A', then the following equation holds good

$$A^2\delta_2(\delta_1 + \delta_2) = A'^2(\delta_2 + \delta_2')(\delta_1 + \delta_2 + \delta_2'),$$

or if we put X for $\delta_1 + \delta_2$ and X' for $\delta_1 + \delta_2 + \delta_2'$, we may write it in the form

$$\delta_2 = \frac{X'\delta_2'}{\left(\dfrac{A}{A'}\right)^2 X - X'}$$

But $\qquad X = \delta_1 + \delta_2$

therefore $\qquad \delta_1 = X - \dfrac{X'\delta_2'}{\left(\dfrac{A}{A'}\right)^2 X - X'}$

where $\qquad X = 3\cdot1416\,x\dfrac{y}{\sqrt{1 - y^2}}$

To determine δ_1 we have therefore to take two resonance curves, one as above described, and another in which the circuit of the cymometer has its decrement increased by a known amount, by the insertion of a second fine wire resistance in the gap provided for it.

The details of the measurements will perhaps best be understood by going through the calculations in a particular case.

A certain oscillation circuit was set up, and by means of the cymometer a pair of resonance curves drawn, one without and one with an added small resistance in the cymometer circuit. These curves were as shown in Fig. 16. From these curves measurements were made giving us the R.M.S. values of the currents a, A, a', A', and at the same time of the quantities $x = 1 - \dfrac{n_2}{n_1}$ and $y = \dfrac{a}{A}$. A number of values of y were taken off the curve corresponding to various values of x not exceeding 0·05, and tabulated as under, and the value of $\delta_1 + \delta_2$ calculated by the formula above given.

$\dfrac{a}{A} = y$	$\left(1 - \dfrac{n_2}{n_1}\right) = x$	$X = \delta_1 + \delta_2$
0·95	0·0120	0·115
0·90	0·0165	0·112
0·85	0·0205	0·104
0·80	0·0255	0·107
0·75	0·0298	0·105
0·70	0·0335	0·103

The mean value of X is then 0·108.

In the same manner, after increasing the resistance of the cymometer circuit, a second set of values was obtained as follows:

$\dfrac{a'}{A} = y$	$\left(1 - \dfrac{n_2}{n_1}\right) = x$	$X' = \delta_1 + \delta_2 + \delta_2'$
0·95	0·0125	0·120
0·90	0·0210	0·138
0·85	0·0255	0·130
0·80	0·0300	0·125
0·75	0·0345	0·124
0·70	0·0385	0·119

Hence the mean value of X' is 0·126.

Accordingly we have from the curves and formulæ above given

$$\left(\frac{A}{A'}\right)^2 = 2\cdot34$$

$$\delta_1 + \delta_2 = 0\cdot108$$
$$\delta_1 + \delta_2 + \delta_2' = 0\cdot126$$

Hence
$$\delta_2' = 0\cdot018$$
$$\delta_2 = 0\cdot017$$
$$\delta_1 = 0\cdot091$$

The greater part of the decrement δ_2 is due to the resistance of the fine wire thermo-junction, and apart from this the decrement of the cymometer in itself is only $0\cdot005$. In this case the oscillation circuit being tested comprised a condenser or Leyden jar and an inductance of 5000 cms. and a spark gap of 2 or 3 mm. in length. The high frequency resistance of the inductance was calculated from the dimensions of the wire and found to be $0\cdot23$ ohm. As this circuit was a nearly closed circuit, the decrement was all due to resistance, partly of the metallic wire R and partly of the spark r, and this can be shown to be equal to $4n_1L\delta_1$, where L is the inductance of the circuit and n_1 the frequency corresponding to resonance. Hence, if R and r are measured in ohms and L in centimetres, we have

$$R + r = \frac{4n_1L\delta_1}{10^9}$$

But $R = 0\cdot23$, $L = 5000$, $\delta_1 = 0\cdot091$, and $n_1 = 0\cdot95 \times 10^6$. Hence $r = 1\cdot23$ ohms.

Also from the formula $M = \dfrac{4\cdot605 + \delta_1}{\delta_1}$ we can show that each train of oscillations comprised about 50 semi-oscillations, or 25 periods.

Accordingly, the measurement of the decrement gives us all information about the nature of the oscillations taking place and the resistance of the spark.

If we had been testing the decrement of a radiotelegraphic antenna, we should have found a much larger decrement than $0\cdot091$, because then there would have been radiation to increase the damping, and therefore the decrement.

It will be seen, therefore, that by the use of the cymometer and the necessary adjuncts to it, we are enabled to obtain all the required information concerning the oscillations in the antenna of a radiotelegraphic transmitter employing the spark method of producing damped oscillations. When operating as above upon an antenna which is inductively coupled to the condenser circuit, the resonance curves will be found to be curves with double humps, as in Fig. 16, Chapter I.; and if these humps are not too close to one another, we may apply the above process to each hump separately, and obtain the decrement of each of the two co-existing oscillations in the antenna.

In making these measurements, the cymometer must of course stand on a table, and a certain length of the antenna must be bent round so as to be parallel with, but not too near, the bar of the cymometer. It will also be found necessary that the outer tube of the condenser should be connected to the earth by means of a terminal provided for that purpose.

9. **Measurement of High Frequency Resistance.**—It has been already explained in Chapter I. that the resistance of a wire of high frequency currents to electric oscillations may be greater than its resistance to ordinary or steady currents by an amount depending on the size and material of the wire. Since the wires generally used for conveying oscillations are round copper wires, we can, in general, by the help of the formulæ given in Chapter I., predetermine the resistance of such circuits for oscillations of known frequency, provided that the circuit consists only of a single wire having a slight curvature. If, however, the wire is in the form of a helix, or otherwise closely coiled, there is no way of determining the high frequency resistance except by enclosing the circuit in the bulb of an air thermometer, and determining the heat produced in it by oscillations of known mean-square value. If, however, we construct the circuit of fine silk- or cotton-covered wires twisted together, each one not having a larger diameter than No. 36 S.W.G., then we prevent the change in distribution over the cross-section of the conductor, and so prevent this conductor from having a different resistance to electric oscillations from that which it has for continuous currents. It is therefore very important that the coils of all oscillation transformers and circuits used in radiotelegraphy should not be formed of solid metal wires, or stranded cables with individual wires of large diameter, but should be formed of stranded cables constructed of very fine insulated copper wires. In this manner we may generally arrange to avoid having to consider the increase in resistance of our metallic conductors to electric oscillations, and from the known or steady current resistance we can calculate that part of the decrement which is due to the resistance of the wire, since it can be shown that in all cases the decrement per half-period due to resistance is equal to the quotient of the resistance of the circuit

divided by four times the produce of the frequency of the oscillations and the inductance, provided that the inductance and the resistance are both measured in consistent units, that is to say, if the resistance is measured in ohms the inductance must be measured in henrys.

On the other hand, when we are concerned with oscillatory circuits in which we have a spark gap, part of this resistance is due to the resistance of the electric spark itself, and this is very variable, depending upon the quantity of electricity conveyed by the spark, the spark length, and also the number of sparks per second. Many measurements have been made of the resistance of electric sparks, but some of these are useless to the radiotelegraphist, because they are concerned only with the resistance of single sparks. Two methods have been adopted for measuring the spark resistance, which lead to different results. In one of these we measure the resistance of sparks of various lengths conveying always the same quantity of electricity, and by the other method we measure the resistance of sparks of various lengths conveying different quantities of electricity.

The first method has been employed by Slaby, the author, and others, and it consists in forming an oscillatory circuit with two spark gaps in it, one of which is variable in length. The circuit includes a hot wire ammeter for measuring the mean-square value of the oscillations, and the experiment consists in substituting for one of the spark gaps a conductive resistance of variable amount, adjusting this latter until the mean-square value of the current in the circuit is the same both for the conductive resistance and when the place is taken by a spark gap of known length. Generally speaking, this method, however, has only an academic interest, because in most circuits as used in radiotelegraphy, the quantity of electricity passing will vary with the length of the spark. Thus, for instance, if we form an oscillatory circuit comprising a condenser, inductance, and spark gap, charging the condenser by means of an induction coil or transformer, then the quantity of electricity put into the condenser depends upon the spark length, because this determines the maximum voltage; hence, when the spark happens, and the oscillations take place, the quantity of electricity that passes through the spark gap at each oscillation is a function of the spark length. The only way in which the spark resistance in these cases can be measured is by means of the cymometer or equivalent process for determining the total decrement of the circuit. If the radiation is absent or very small, then the total decrement is made up of two parts, a part depending upon the high frequency resistance of the circuit and a part depending upon the resistance of the spark. If we call d_1 the part of the decrement due to the high frequency resistance, and d_2 that part of the decrement due to the spark resistance, and if we call R the high frequency resistance of the metallic part of the circuit and r the resistance of the spark, both measured in ohms, then, provided there is no source of loss of energy in the condenser itself, and no loss of energy by radiation, the sum of these two decrements is connected with the sum of these two resistances by the formula

$$d_1 + d_2 = \frac{(R + r)10^9}{4nL}$$

where L is the high frequency inductance of the circuit in centi- metres and n is the frequency of the oscillations.

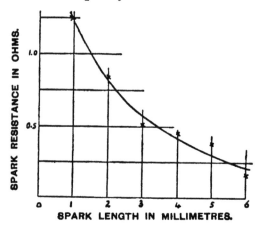

FIG. 17.

Hence, if we can determine by the cymometer the frequency of the circuit, and determine by calculation the high frequency resistance and inductance of the metallic part of the circuit, and determine experimentally, as already described, the total decre- ment, we are able to calculate the spark resistance. In this manner it can be shown that the spark resistance gradually decreases with increasing length of spark, reaching after a time a nearly constant minimum value, and that it varies to some extent with the materials of which the spark balls are made and with the gap in which the spark takes place.

For very large spark discharges this resistance will be only a small fraction of an ohm, but for short sparks two or three milli- metres in length, such as take place when Leyden jars are charged by an induction coil, as in short distance radiotelegraphic apparatus, the spark resistance may amount to several ohms.

The curve shown in Fig. 17 embodies the results of observations by the author on the spark resistance of sparks of various lengths taken in the above manner ; the condenser in the circuit consisting of metallic plates immersed in oil, and the inductance a single rectangle of fine wire, the high frequency resistance and inductance of which were calculated from its dimensions. The capacity used was 0·00261 mfd., and the inductance 0·00636 millihenry.

CHAPTER IX

1. The Problem of Radiotelephony.—Before the invention of the methods of radiotelephony described in this chapter, attempts had been made with some degree of success to transmit the sound of articulate speech over moderate distances without the aid of a connecting wire. In addition to a method depending upon the induction of currents and their conduction through the earth, another has been worked out based upon a peculiar property of selenium of varying its resistance under the action of light and of the continuous current electric arc of varying the intensity of its light when a periodic current is superposed upon the continuous one. We shall, however, here confine our attention to the details of the method employing electromagnetic waves. which gives the greatest promise of ultimate utility, now generally called Radiotelephony.

Radiotelephony consists, therefore, in the transmission to a distance of articulate speech through space without wires by means of electromagnetic waves, as distinguished from radiotelegraphy, which is the transmission of intelligence by means of arbitrary signs, whether audible or visible. As soon as radiotelegraphy, as conducted by the methods already described, had made a certain progress, inventors had their minds naturally turned to the problem of the transmission of articulate speech by the same means. It very soon, however, became clear that the attainment of any practical success was bound up with the invention of a transmitter for producing undamped electric radiation, and of a receiver which should be quantitative in action, that is to say, not merely set in operation by oscillations, but produce an effect proportional to the amplitude of the waves incident on the receiving antenna.

The oscillation detector to be used in connection with radiotelephony must therefore be of such a character that it is capable of varying the current through a telephonic receiver in exact correspondence with the variations of air pressure due to the speaking voice taking place in proximity to the particular telephonic transmitter employed at the sending station.

In electric telephony conducted with wires, the apparatus usually employed consists of a transmitter of the microphone type and a receiver of the magnetic or Bell type. For instance, in the simplest form of short distance transmitter and receiver the microphone transmitter consists of a metal diaphragm which is set in vibration by the variations of pressure taking place in proximity to the mouth of a speaker uttering near it articulate words. Behind the diaphragm is some arrangement by which a variable or imperfect contact between carbon surfaces is altered by pressure. In the ordinary type of granular carbon microphone, the movements of the diaphragm are made to press together more or less small fragments of graphitic carbon contained in a shallow chamber, and so alter the electric conductivity of the mass. This

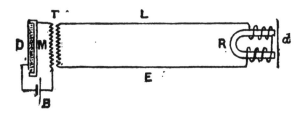

Fig. 1.

variable carbon resistance, M, is placed in series with a few voltaic cells, B, and the primary circuit of a small induction coil, T. One end of the secondary circuit of the induction coil is connected to the earth or to one of the line wires, L, and the other to the line wire or duplicate line wire, if a complete metallic circuit is employed (see Fig. 1). At the receiving end the currents in the line pass through the magnetising coils, R, of a magneto-telephone consisting of a permanent magnet having its polar extremities surrounded with these magnetising coils, the pole or poles being placed in proximity to a thin sheet iron diaphragm, d. The motions of the transmitting diaphragm, D, are therefore repeated by the receiver diaphragm, and every sound made near the microphone transmitter is reproduced by the diaphragm of the receiver. If, for instance, a musical sound is created near the diaphragm of the transmitter, there will be variations of air pressure which may be represented by the ordinates of a periodic curve. In the case of a perfectly pure musical sound, this curve approximates in form to a sine curve, but for any such sound as a vowel sound, the form of the curve will be complicated, although periodic if the vowel sound is continued. By various devices it is possible to delineate graphically the periodic curves corresponding to various musical or prolonged vowel sounds as in the diagrams in Fig. 2, which are the results of experiments by Mr. W. Duddell, who has kindly given permission to reproduce them here. In the case of articulate sounds, the variations in air pressure are non-repetitive, but they can nevertheless be represented by the ordinates of a curve. Thus, for instance, in speaking to a gramophone or phonograph, the voice creates variations of air pressure in front of the speaking diaphragm, and at the back of this diaphragm, or connected with it by a system of levers, is a delicate cutting tool which carves out upon the surface of the moving plastic cylinder or disc which forms the receiving surface a little channel or groove, the bottom of which is irregular, the depth of this groove corresponding from instant to instant to the variations of pressure produced against the diaphragm by the speech being made. If, therefore, a section could be made of this groove, and the outline of the bottom enlarged, it would present the appearance of a very irregular non-repetitive curve, each change in the ordinate of which, however, corresponds to a change in air pressure of the air in front of the diaphragm against which the speech is being uttered, and has therefore a signification as far as the ear is concerned.

The problem of telephony is therefore to cause some other diaphragm at a distance to be moved from instant to instant in a similar manner to that of the diaphragm against which speech is being made. This receiving diaphragm will then reproduce at the distant end the same variations of air pressure as those which actuated the transmitting diaphragm, and a human ear placed in proximity to it will therefore hear the speech being made at the distant place.

In telephony with wires, the movements of the transmitting diaphragm are made to translate themselves into corresponding variations in the strength of an electric current in the connecting wire by means of the variation in resistance which takes place when carbon surfaces are more or less pressed together, and the re-translation of this variable electric current into the movement of a receiving diaphragm is made to take place by means of the variations in the polar strength of a permanent magnet which takes place when an electric current of varying strength circulates round that pole.

To achieve radiotelephony we remove the interconnecting wire

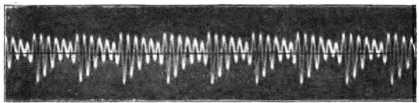

Vowel ā as in Ma.

Simple Form of ōō Sound as in Coo.

Complex Form of ōō Sound as in Coo.

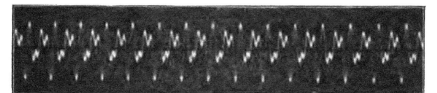

Vowel ō as in Ho.

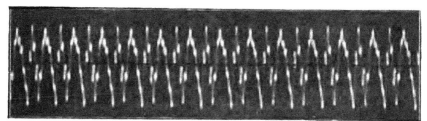

Vowel ē as in Me.

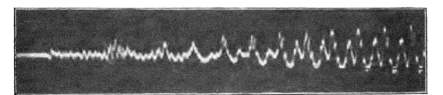

and substitute for it a train of electromagnetic waves passing through space. Hence the particular inventions required in order to accomplish the desired result, as regards the transmitter, are to devise a mechanism capable of emitting undamped electromagnetic waves and to vary the amplitude of these waves in accordance with the variations in the air pressure taking place against the transmitting diaphragm, and at the receiving end to cause these undamped waves of variable amplitude to actuate a mechanism which shall cause them to set in vibration the receiving diaphragm, so that its displacements correspond with the variations in the amplitude of the electromagnetic waves.

We have therefore to consider (1st) the transmitting arrangements which have been invented, and (2nd) the receiving arrangements in connection with radiotelephony.

2. **Transmitting Arrangements in Radiotelephony. High Frequency Alternator Method.**—It is generally agreed that for the perfect transmission of articulate speech by electromagnetic waves an essential condition is the possession of means for producing at the transmitting station, in the sending antenna, undamped or practically undamped electric oscillations. We have at the present time available several effective means of doing this, such as the high frequency alternator, or by the use of a carbon-metal continuous current arc in a hydrocarbon atmosphere, as already described in Chapter III.

The alternator method has been particularly advocated and employed by R. A. Fessenden, and the arc method by V. Poulsen and E. Ruhmer.

An essential condition of success in the transmission of articulate speech by electromagnetic waves is that there shall be no interruptions in the uniform flow of the undamped oscillations, at least not below such a frequency as is equal to the upper limit of the frequency of audible sounds. If regular vibrations are set up in the air, these are appreciated as sound by the normal ear if they lie in frequency between 40 and 20,000 per second. Human ears vary, however, a great deal in the value of the highest frequency which can be heard as a sound. As regards musical sounds, the highest frequency employed does not exceed 4000 or 5000. If intermittent trains of damped waves were employed, even if the frequency of the trains was as much as 4000 or 5000, they would affect the oscillation detector at the receiving station, and hence the telephone in connection with it, and produce in the latter a musical sound of high pitch which would drown out the variations of lesser frequency which constitute the articulate speech.

Hence, if an alternator producing an alternating current having a frequency of even 10,000 were connected to a radiating antenna, it is probable that most persons would hear a sound in a telephone connected to an electrolytic oscillation detector in a receiving antenna. If, however, it had a frequency of 20,000, they would probably not hear any sound. We may say, therefore, that to be of practical use in radiotelephony a high frequency alternator should give a current having a frequency of not less than 20,000, and preferably higher. Fessenden has constructed such an alternator of the Mordey type, as already mentioned, with fixed armature and revolving field, having 360 poles or teeth. At a speed of 139 revolutions per second it gave a terminal E.M.F. of 65 volts, and an alternating current with a frequency of 50,000, the maximum output at this frequency being 300 watts. The attainment of a speed of 8000 R.P.M. in any revolving shaft and disc necessitates very perfect balancing, and if used on board ship special devices are necessary to obviate gyrostatic action, which would cause serious wrenching at the bearings as the ship pitches or rolls.

As regards driving power, one form of motor adopted has been the De Laval steam turbine, which, in small sizes, can be made to run up to 30,000 R.P.M., or 500 revolutions per second. Electric motors have been constructed, however, for the author, which for 1-H.P. size have run up to 6000 R.P.M., and for 5-H.P. size up to 4000 R.P.M., and these speeds can be multiplied by the employment of a thin and very flexible belt.

The attainment of high speeds is facilitated by the use of ball or cylinder bearings and adequate lubrication, and, above all things, by perfect balancing.

In the general design of the alternator, the choice lies between the inductor type of machine in which the only revolving part is an iron disc with teeth cut on the edge and the type of alternator with wound polar teeth. The inductor type of alternator has the advantage that it is easy to balance it, but, generally speaking, there is a considerable decrease in terminal voltage with increase in current taken out of the machine. Fessenden has pointed out that it is important that a high frequency alternator should give an electromotive force curve having a true sine form, as it is only then that resonance can be employed to multiply the electromotive force. For this reason it is necessary that there should be no iron in the armature, and as the polar teeth of the field magnet must be very narrow and close together, not more than two or three millimetres in width, to secure the necessary frequency with available speeds of rotation, it is essential to place all the teeth of one polarity on the same side to avoid magnetic leakage. This then results in the selection of the

Mordey type of alternator with fixed non-iron armature and revolving fields.

It is not necessary to generate very high E.M.F., as we can always transform it up by an oscillation transformer or by resonance, but it is necessary to have high frequency.

R. A. Fessenden (see *The Electrician*, Vol. 61, p. 441, 1908) has constructed a high frequency alternator direct coupled to a De Laval turbine, giving a current at 225 volts, with a frequency of 75,000 and about 2·5 kw. output. The machine is of the double armature type, with 300 coils on each, and a field with 150 teeth. The two air-gaps are only $\frac{1}{16}$ inch in length (see Fig. 3). The required steam pressure is 100 lbs. on sq. inch.

[Reproduced by permission from " The Electrician."

FIG. 3.—High Frequency Turbo-Alternator (Fessenden).

Also Alexanderson has constructed high frequency inductor alternators as described in Chapter III.

A disadvantage of such machines is that they are non-portable and not suitable for use on board ship. Nevertheless, in land stations it would appear to provide a possible form of generator for long-distance radiotelephony.

3. **Electric Arc Transmitters for Radiotelephony.**—Turning then to the other method already described for the production of undamped oscillations by means of the electric arc in a hydrocarbon atmosphere, we find that V. Poulsen has brought such arc oscillation generators to a considerable degree of perfection, and

applied them also very successfully in the transmission of speech for very considerable distances over land and sea.

As the method of producing undamped oscillations by the aid of a continuous current arc formed between a rotating carbon cathode and a cooled copper anode in hydrocarbon gas has already been fully described in Chapters III. and VII., we need only here consider the most recent modification of the apparatus. To get rid of the necessity for cooling the box containing the electric arc with water, it is constructed with radiator flanges, so as to be air cooled, and it has been found that alcohol vapour is a convenient substitute for

[*Reproduced from " Electrical Engineering," of April 23, 1908, by permission of the Proprietors.*

Fig. 4.

coal gas or hydrogen, so that methylated spirit is admitted drop by drop into the chamber in which the arc burns by a form of sight-feed lubricator controlled by a hand or electro-magnetic valve (see Fig. 4). By these improvements the arc method has been made independent of the supply of water and coal gas which was necessary in the early forms of apparatus. A small arc can also now be employed using 200 or 220 volts and a current of 1·5 or 2 amperes. It therefore becomes quite easy to operate it off any commercial lighting circuit furnishing continuous current at 220 volts, and its power consumption is not more than 300 or 400

watts.   With this small power consumption troubles do not arise
from the deposit of soot in the arc chamber.   The current through
the arc is regulated by adjusting the distance of the carbon and
copper electrodes by a screw which is hand regulated, that is to
say, an assistant keeps the current through the arc constant by
watching an ammeter in series with it and adjusting the screw.
The arc is formed in a powerful transverse magnetic field, and the
carbon is made to rotate slowly by means of clockwork or a geared-
down electric motor.   Connected to the copper and carbon
electrodes of the arc is an oscillation circuit, consisting of a
variable inductance formed of a helix of wire, and a condenser of

[*Reproduced by permission of the Amalgamated Radiotelegraphic Co.*

FIG. 5.

variable capacity consisting of metal plates in oil.   The form of
condenser preferred is one in which there are a number of semi-
circular plates fixed one above the other in a tall cylinder of highly
insulating oil (see Fig. 5).   A number of other semi-circular plates
sandwiched in between the first-named set are affixed to a long
metal rod, so that by turning this rod round the movable plates are
brought more or less in between the fixed set of plates.   The two
sets of plates constitute the surfaces of the condenser, and the oil
with which the jar is filled the dielectric.   Hence, by simply
turning round a milled head on the top of the cylinder, large
variations of capacity can be created.   The variable inductance

joined in series with this condenser is generally a single helix of bare wire with a sliding contact, by means of which more or less inductance can be introduced. In constructing the oscillation circuit to be used as a shunt across a continuous current arc, it seems important that the capacity should be kept small and the inductance large. That is to say, that a given oscillation constant for that circuit should be obtained not by the use of a small inductance and a large capacity, but by the use of a large inductance and a small capacity. If the capacity is reckoned in electrostatic units, and the inductance is reckoned in centimetres, then the capacity so reckoned may be to the inductance in the ratio, say, of 1 to 20.

This oscillation circuit is coupled inductively to an antenna circuit, the coupling being close, and the antenna circuit syntonised with the condenser circuit by the introduction of suitable inductance, so that oscillations set up in the condenser circuit induce others of equal frequency and a maximum strength in the antenna circuit (see Fig. 6). Generally speaking, the current in the oscillation circuit will be a current of 4 or 5 amperes, R.M.S. value, and the frequency may be anything between 100,000 and a million. When the arc is set in operation we have undamped oscillations created in the condenser circuit and undamped waves emitted by the antenna.

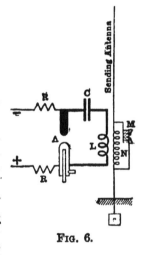

Fig. 6.

4. **Microphonic Control of Electric Oscillations.**—In order to conduct radiotelephony, we have then to control the amplitude of the electric waves emitted by the antenna, so that this amplitude may vary in exactly the same manner and proportionately to the change of air pressure at any point near the mouth of the person uttering articulate speech. This is best done by the insertion of a speaking microphone in the condenser shunt circuit. Such a speaking microphone consists of a shallow metal chamber closed by a flexible metal diaphragm, which is insulated from the metal chamber, the space between the diaphragm and the solid back containing carbon granules which are more or less compressed by the vibrations of the diaphragm. Hence, when speech is made against a mouthpiece terminating on the diaphragm, the aerial vibration sets up similar vibrations in the diaphragm, and this movement, by compressing more or less the carbon granules, varies the resistance of the carbon included between the diaphragm

z

and the solid back. As a single microphone transmitter cannot be operated satisfactorily with a large current, when it is desired to introduce it into a circuit in which a current of 4 or 5 amperes is flowing, it is necessary to employ a number of these microphone transmitters arranged in parallel, so that each microphone may not carry more than 1 ampere. The microphones can be arranged in a box or at the ends of branching pipes, so that they are all simultaneously affected by variations of air pressure due to speech made to a single mouthpiece (see Fig. 7).

In the arrangements adopted by Poulsen this microphone transmitter or variable resistance is inserted either in the condenser

[*Reproduced from " Electrical Engineering " of April 23, 1908, by permission of the Proprietors*

FIG. 7.

shunt circuit of the arc or else in a tertiary circuit closely adjacent thereto (see Fig. 8). When speech is uttered against the microphones it varies the resistance of this microphone circuit, and therefore alters the resistance of the condenser circuit slightly, and therefore also affects the current in the sending antenna. Words spoken to the mouthpiece, therefore, produce an effect upon the amplitude of the emitted electric waves, and these are, so to speak, moulded into speech form, that is to say, made to vary as the ordinates of a wave curve representing the changes of air pressure taking place in the mouthpiece of the transmitter.

In some cases the microphone resistance may be inserted in

the circuit of the electric arc and operate directly upon the continuous current affecting the arc. In this case, a variation of the condenser current and also of the amplitude of the wave radiated from the antenna takes place in the same manner as the variations in the arc current produced by the changes in resistance of the microphone under the action of the articulate sounds. Or again, the microphone may be inserted as a shunt to the secondary circuit of the oscillation transformer connecting the antenna to the condenser circuit, so that the current into the antenna is more or less shunted to earth (see Fig. 6). Finally, the microphone may be inserted in the earth connection of the antenna so as to vary the current flowing into the antenna itself, and therefore the intensity of the radiated waves. In any case, it should be inserted at a node of potential in the oscillatory circuit.

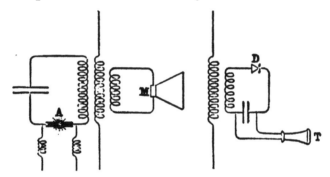

[*Reproduced from " Electrical Engineering " of April 23, 1908, by permission of the Proprietors.*

FIG. 8.—Scheme of Circuits of Transmitter and Receiver for
Radiotelephony.

Another plan that has been suggested is to employ a condenser telephone consisting of two plates of metal near together, one of which constitutes the diaphragm against which speech is made, so that by the vibrations of this diaphragm under the operation of the voice the plates are more or less approximated, and their electrostatic capacity varied. If this condenser transmitter is joined in parallel with the main condenser in the oscillation circuit, speech made against it, by altering its capacity, more or less throws the condenser circuit out of tune with the antenna circuit, and therefore varies the intensity of the emitted waves. By any of these methods the train of undamped waves emitted by the antenna may be moulded into the form of speech, and these waves are then retranslated back into sound by the arrangements employed in the receiving circuit.

The picture in Fig. 9 shows the complete arrangement for employing a Poulsen arc in a radiotelephonic transmitter apparatus. Fig. 7 shows the multiple microphone arrangement, the mouthpiece against which speech is made branching out into a number of parallel pipes, at the end of each of which is a carbon microphone made as described, such a transmitter being suitable for radiotelephony over 250 miles.

By these methods Poulsen has succeeded in transmitting articulate speech from Berlin to Copenhagen, a distance of 460 kilometres, or 290 miles.

When a high frequency alternator is used as the source of the

[*Reproduced from " The Electrician " by permission of the Proprie'ors.*

FIG. 9.—General View of the Poulsen Transmitting Apparatus for Radiotelephony.

undamped oscillations in the sending antenna, the oscillations created by the alternator are transferred to the antenna through an oscillation transformer. The microphone may then be placed in the antenna circuit, and as near the earth as possible, or it may be placed in a tertiary circuit wound on the oscillation transformer itself.

5. Other Arrangements employed as Transmitters in Radiotelephony.—To avoid placing the arc in a strong magnetic field and enclosing it in an atmosphere of hydrogen or hydrocarbon, experiments have been made with a number of electric arcs in series. It has already been explained, in Chapter III., that the characteristic curve of a continuous current arc is steeper for small currents than for larger ones. If, then, a number of electric arcs taking a small current are joined in series, and a condenser circuit shunted over the whole series, it is possible to arrange these arcs to take a small current and obtain from them high frequency oscillations.

This has been carried out as follows: A copper tube has a curved concave bottom fixed into it, and this is filled with water to keep it cool. This forms the positive terminal of the arc. The negative terminal is formed by a solid carbon rod, and the arc is struck in the cavity formed by the recessed end of the copper tube.

Six such arcs are joined in series and a condenser and inductance shunted over the whole series.

It is easy to contrive a mechanism by which all the arcs shall be struck at once and controlled together. The arcs are arranged to take a very small current. The oscillatory current set up in the condenser circuit is made to act inductively upon a syntonic antenna, and a microphone or number of microphones are used in parallel to regulate the emitted waves into speech form. This microphone may be placed in the antenna circuit or as a shunt to it.

Many modifications of the arc generator have been devised for the purposes of radiotelephony. Also methods of generating the undamped currents required by means of quenched sparks of extremely high spark-frequency have been applied.

Thus, in France Colin and Jeance have done much work on radiotelephony using as arc electrodes thin carbon discs in an atmosphere of acetylene and hydrogen in certain proportions, which reduces the wear of the carbons to practically nothing. Also they dispense with the magnetic field which requires a cumbersome electromagnet.

Again, in America Janke has employed an electric arc under liquid alcohol. Generally speaking, however, both the alternator and arc methods of generation labour under serious objections.

The high frequency alternator is an expensive, heavy, and complicated machine which requires a supply of electric power to drive it, and even at best furnishes a current having a frequency corresponding to a very large wave length. Thus, a frequency of 50,000 corresponds to a wave length of 20,000 feet, and this is not suitable for small antenna.

On the other hand, the arc method is by no means free from difficulty. It is an inefficient method of creating high frequency currents. It is also slow in starting, and requires to run for some time to get into a steady condition. Also it is unstable and upset by sudden variations of load, and does not furnish absolutely unbroken undamped currents.

Hence wireless telephony by means of an arc generator has always been of the nature of an experimental feat and cannot be said to have reached a condition in which it can be used in practice like radiotelegraphy on the spark system.

Attempts have, therefore, been made to develop other more reliable generators. Three Japanese inventors, Messrs. Torikata, Yokoyama, and Kitamura, have brought out a generator called the T.Y.K. system from the initials of their names. In this generator a direct current of about 0·2 ampere under an electromotive force of about 500 volts is passed between electrodes of magnetite (oxide of iron) and brass. These electrodes are small, flat surfaces of about 1 sq. cm. in area. The distance between them is controlled by electromagnetic regulator something like the control of an arc lamp. These electrodes are shunted by a circuit having capacity about 0·05 and inductance, and in this circuit high frequency oscillations are created. The inventors consider that the discharge between the electrodes is of the nature of a very rapid series of sparks rather than a true arc. The oscillations in the condenser circuit are made to induce others in the transmitting antenna, and in the base of this last a solid back carbon microphone is placed. The inventors use a current of about 1 ampere in the transmitting antenna which can pass without much heating through a single microphone. The reception is conducted by means of a " perikon " or zincite-chalcopyrite rectifying detector in series with an ordinary double head telephone. The inventors claim that it is quite a practical system, usable by any one up to 10 or 15 miles, and by experts to a much greater distance.

Another quite different and ingenious form of undamped generator has been developed by Mr. H. J. Round and the experts of the Marconi Wireless Telegraph Company.

The generator is based on the properties of the valve glow lamp originally invented by the author as a receiver. In the author's oscillation valve a metal plate or cylinder carried on a separate sealed-in platinum wire is included in the bulb of an incandescent lamp, either carbon or metallic filament. When the filament is incandescent there is a projection of negative electrons from it which bestows upon the vacuous space a unilateral conductivity. Suppose then that such glow lamp has not only a plate but also perforated grid sealed into the bulb, and that connections with batteries, transformers, and telephones are made as in Fig. 10.

In this diagram T_1 and T_2 are two telephones or a telephone and a microphone transmitter. C is a battery for igniting the lamp filament F, and D is a battery for varying the potential of the grid, and V is another high potential battery.

Then consider what happens if the plate H is made positive and the grid G is kept at zero potential. Then electrons will pass from the filament to the plate, and there will be a current flowing through the primary of the transformer B. If then the potential of the grid G is varied + or − ever so little, it will vary very much the number of electrons which reach H, provided that we are working on the steep part of the characteristic curve of the valve. Hence, if speech is made to the telephone T_1 the effect on a receiver at T_2 will be very much greater than if T_1 and T_2 were directly connected. In other words, the arrangement will act as a magnifier of telephonic speech. It is well known that if a microphone, battery, and telephone receiver are connected in series that when the receiver is approached to the transmitter the arrangement begins to sing spontaneously and emits a musical note.

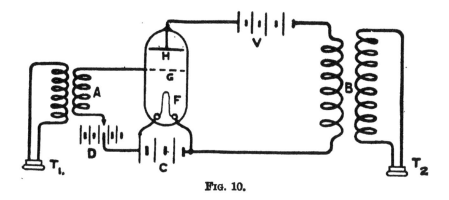

FIG. 10.

The explanation of this is as follows : Stray sounds start the diaphragm of the transmitter in vibration. This starts the receiver diaphragm vibrating and it emits a sound. This sound increases the amplitude of the transmitter vibrations, and so the two act and react on each other until a loud sound is emitted by the receiver.

Returning, then, to the arrangement in Fig. 10, if the two transformers are connected together, as suggested by Messner in 1913, the arrangement will act like a coupled microphone and telephone, and electric oscillations when once set up will be sustained.

Starting from this fact, Mr. H. J. Round has developed a glow lamp valve generator, for generating perfectly smooth oscillations in a transmitting aerial. These are modulated to speech form by

FIG. 11.—Transmitting and Receiving Apparatus for Wireless Telephony, based on the use of the Valve Generator and Receiver (Marconi-Round system).

a single carbon microphone, since the antenna current is mûch less than 1 ampere. As a receiver, Mr. Round employs a vacuum valve magnifier, as above described, associated with a crystal rectifier.

The transmitting valve is a specially contructed large lamp, and the whole arrangement is shown in Fig. 11. The 50 kilometre set, made by the Marconi Wireless Telegraph Company, delivers 0·6 ampere to the sending antenna, and this can be increased to 1 ampere per 100 miles transmission.

The great advantage obtained by the use of the ionised gas magnifier as a receiver, is that we can then obtain effective radiotelephony over considerable distances without the employment of currents in the sending antenna larger than can be modulated by an ordinary carbon microphone.

In those systems which employ a Poulsen arc or alternator as the generator when the sending antenna current reaches 5 or more amperes, the great difficulty is to construct a microphone that will pass and modulate this large current.

For this purpose, Professors Majorana and Vanni in Italy have devised liquid microphones, in which the resistance, which is modulated or affected by the voice, is a stream or film of acidulated water. Again, Marzi has invented a microphone in which a stream of carbon particles falls between two surfaces, one of which is moved by the voice, so that the carbon particles are more or less compressed, but are continually being renewed, so that they cannot become heated.

Lieuts. Colin and Jeance in France have employed this Marzi microphone with success in radiotelephony.

W. Dubilier and others have invented water-cooled carbon microphones capable of being used with fairly large currents.

6. **Receiving Arrangements in Radiotelephony.**—Several forms of oscillation detector already described are very suitable for radiotelephonic reception, such as the Fessenden electrolytic detector, the author's glow lamp or ionised gas detector, the thermoelectric detectors, and the crystal rectifiers. Thus, for instance, if a receiving circuit is constructed by inductively coupling the receiving antenna to another oscillation circuit comprising a condenser, an inductance properly syntonised to the antenna circuit, and also includes an electrolytic detector coupled as already described, to a telephone and a local cell, the oscillations passing through the electrolytic detector will not merely alter its apparent electrical resistance, but alter it in some sense proportionately to their intensity, and hence, if undamped waves are falling upon the antenna of constant wave length but varying amplitude, a variation in the

apparent resistance of the electrolytic detector will take place, which follows and imitates the variation of wave amplitude. Accordingly, the current sent through the telephone by the local cell varies in the same manner, and the telephone diaphragm therefore emits a sound which corresponds with the amplitude of incident electromagnetic waves, and therefore reproduces speech being made against the diaphragm of the transmitting microphone.

Fessenden makes use of a form of telephone receiver he calls a "heterodyne" receiver. It consists of a pair of coils of wire, one of which is wound round an iron wire core, and the other is attached to a mica diaphragm held near the core. The last-named coil is traversed by the current in the receiving antenna, and the first by a local current of the same frequency as that of the transmitter. There is therefore a mechanical force between the two coils which varies with every variation of the current in the receiving antenna. The diaphragm therefore reproduces the sounds which are made against the diaphragm of the microphone in the transmitting circuit.

We have, therefore, in the combined radiotelephonic transmitter and receiver, a wonderful transformation of energy. The variations of air pressure made against the speaking diaphragm produce similar variations in the resistance of the microphone; this again varies in the same manner the intensity of the electric oscillations set up in the oscillation circuit connected with the arc, and also the oscillations in the antenna. Thus electromagnetic waves are emitted, the amplitude of which is changing in the same manner. A portion of the energy of these waves is then translated back by the receiving antenna into oscillations, the amplitude of which varies also in the same manner as that of the incident waves, and these, acting on the particular detector coupled to the telephone, reproduce movements of the receiving telephone diaphragm which imitate those made by the diaphragm of the transmitting microphone.

Although this operation is complicated, yet, nevertheless, it has been so far perfected that articulate speech can now be transmitted several hundred miles by these means. In fact, radiotelephony, or telephoning without wires, seems to have certain undoubted advantages over telephony conducted with wires. It is well known that the reason why telephonic speech cannot be transmitted more than a certain moderate distance through submarine cables, is because of the *distortion* in the wave form which takes place owing to the combined action of the capacity, inductance, resistance, and leakage of the cable. The reason for this is that electrical vibrations of different frequency travel through such a cable with different velocities. Hence, when a complex vibration is impressed upon the cable by means of an ordinary telephone transmitter, the complicated wave form which represents, as already explained, any spoken word, can be resolved into the sum of a number of simple or sinoidal or harmonic vibrations of different frequency. These vibrations travel through the cable at unequal rates, and hence beyond a certain distance the integral wave form is distorted beyond recognition by the ear. This distortion may be compared with that of bad handwriting. In the case of ordinary written words, a single letter is hardly ever perfectly formed; but if the departure from perfect writing does not exceed a certain limit, our experience enables us to guess pretty quickly what the word really signifies. In the same way, when listening through a telephone, if the distortion of sound does not exceed a certain limit, the ear is able to guess the meaning of the word, but beyond a certain point it is unrecognisable. Radiotelephony, therefore, seems marked out specially for the transmission of articulate speech over sea, owing to the greater difficulty of telephoning through submarine cables than through land wires. Moreover, there is, of course, no necessity that the transmitting and receiving station should remain fixed in position.

Another form of oscillation detector suitable for telephonic reception is the oscillation valve or glow lamp detector, invented by the author, which has been made use of by Mr. Lee de Forest under the name of an audion. This glow lamp detector has already been fully described in Chapter VI. It consists of a carbon filament glow lamp having an insulated metal plate or cylinder in the bulb carried on an insulated terminal. When the carbon filament is incandescent by an insulated battery it emits negative ions, and a current of negative electricity can pass across from the filament to the insulated plate sealed into the lamp bulb, which varies with the voltage between the negative terminal of the lamp and the insulated plate. If, therefore, a telephone is connected between one end of the filament and the insulated plate, as described and shown in Figs. 20 and 21 of Chapter VI, the oscillations produced in the receiving antenna, varying from moment to moment in strength with the amplitude of the incident waves, will send through the telephone a continuous current which also varies in the same manner, and the telephone therefore reproduces the articulate sounds made against the microphone of the transmitting station.

Another form of detector much employed in radiotelephonic work is the crystal rectifier without local battery. In the arrangements employed by Poulsen, the antenna and the earth wires are connected to the two terminals of a condenser shown fixed up against the wall in Fig. 12, which forms with a variable inductance an oscillation circuit tuned to the period of the antenna. This inductance is loosely coupled to another oscillation circuit consisting of a condenser, inductance, telephone, and crystal rectifier. The coupling or mutual inductance of the two circuits is very weak, the primary and secondary helixes frequently being set a considerable distance apart, as shown in Fig. 12. The damped oscillations of constant frequency but varying amplitude taking place in the antenna induce other oscillations of the same period and similarly varying in the telephone and rectifier circuit, and

[*Reproduced from " The Electrician " by permission.*

FIG. 12.—Poulsen Receiving Apparatus for Radiotelephony.

the rectifier permits the current to pass through the telephone only in one direction.

7. **Present State and Achievements of Radiotelephony.**—The transmission of articulate speech to a distance by means of electromagnetic waves without the aid of an interconnecting wire has made remarkable progress in the last few years, and has considerable possibilities of improvement.

Poulsen has succeeded in transmitting phonograph music by this means from Berlin to Copenhagen, a distance of 460 km., or 290 miles. In Fig. 13 is shown the apparatus used by him for this purpose. On the right-hand side of the picture will be seen the transmitting arrangements, consisting of the copper-carbon arc in its box and the inductance and variable condenser of the oscillatory circuit, and on the left-hand side the receiving arrangements, including the receiving telephone. Distinct articulate speech is also said to have been transmitted by the same means

[*Reproduced by permission of the Proprietors of "The Electrician."*]

FIG. 13.—Poulsen's complete Apparatus for Radiotelephony.

from Lyngby to Esbjerg, a distance of 270 km., or 170 miles. The receiver contained a thermoelectric oscillation detector.

Fessenden has also described the arrangements and apparatus of the National Signalling Company of the United States, devised by him for radiotelephonic communication between Brant Rock and New York—350 km., or 200 miles. The generator is a

1 kw. steam turbine-driven alternator, giving alternating currents
of a frequency of 81,700 to 100,000 at 150 volts (see Fig. 3).
The resistance of the disc armature is 6 ohms, and the field
exciting current 5 amperes. Using a transmitting antenna 200
feet high at New York, and the Atlantic Tower, 400 feet high,
at Brant Rock, an energy expenditure of 200 watts in the antenna
is required to cover the 200 miles.

Successful demonstrations were also made in 1906 by the same
inventor between Brant Rock, U.S.A., and Plymouth, Mass., a
distance of 11 miles, in which speech was transmitted, said by
telephone experts' present to be fairly satisfactory.

In 1908 similar experiments were made by Prof. Majorana,
in Italy, between Monte Mario and Porto Danzig, a distance of
60 kilometres, in which good speech transmission was obtained.

In France, Lieuts. Colin and Jeance, and Chief Engineer
Mercier have achieved the distinction of transmitting speech
radiotelephonically from Paris (Eiffel Tower station) to Dieppe,
and musical sounds from Paris to the coast of Finisterre, a distance
of 310 miles.

In Italy, Dr. J. Vanni, using his own liquid microphone
transmitter, has transmitted good speech radiotelephonically for
a distance of over 600 miles between Rome and Tripoli. Successful experiments have been made for the Italian navy by Mr.
Marconi, using the valve transmitter above described.

We may say, therefore, that the transmission of articulate
speech by electric radiation has attained at present (1915) to
something like the range and efficiency reached by radiotelegraphy
ten years ago, and, doubtless, in the next few years will steadily
progress; but much has yet to be done before it can compete
with modern methods of radiotelegraphy in providing regular
communication between ships and the shore. Nevertheless, it is
a most interesting and wonderful application of electrical knowledge, placing at our disposal new means of communication
between distant and even moving stations.

INDEX

A.

B.

2 A

F.

G.

H.

HAMMER interrupter, 48
Henry, Joseph, researches of, on magnetisation by electric discharges, 201
Hertz, Henrich R., 141
Hertz's investigations, 142
Hertzian oscillator, 142
Hertz resonance circuit, 194
High frequency alternators, 81
,,   ,,   ,,   for radiotelephony, 334
,,   ,,   alternating current, definition of a, 2
,,   ,,   resistance of conductors, 9, 324
           ,,   of wires, formulæ for, 15
,,   ,,   ,,   of spiral conductors, experimental investigations
                       on, 17
High speed dischargers of Marconi, 272, 273
Highly damped radiator, 153
   ,,   ,,   train of oscillations, 8

I.

IMPERFECT contact detectors, 194
Inductance, high frequency, 18
   ,,   ,,   ,,   measurement of, 302
   ,,   •   nature of, 19
   ,,   of various circuits, formulæ for the, 21
Inductances for radiotelegraphy, 68
Induction coil, construction of an, 42
   ,,   ,,   qualities required in an, for wireless telegraphy, 44
   ,,   coils, mode of winding secondary circuits of, 43
   ,,   ,,   ,,   of constructing, for wireless telegraphy, 43
Inductive coupling of antenna to oscillatory circuit, 176
   ,,   effects of undamped oscillations, 117
Interrupters for induction coils, 48–51

K.

KLEMENCIC, J., 213

L.

LAW of exchanges, 145
Lee, Miss Alice, 145
Leyden jar, time period of discharge of a, 39
Linear radiator, 142
Locating direction of radial point, experiments on, by Bellini and Tosi, 191
Lodge, Muirhead & Robinson, oscillation detector of, 199
Lodge, Sir Oliver, 196
Lodge's coherer, 196
Logarithmic decrement, definition of the, 7
   ,,   decrement, measurement of, 316
Long-distance radiotelegraphy, 264

M.

N.

O.

P.

Q.

R.

S.

W.

Z.

PRINTED IN GREAT BRITAIN BY WILLIAM CLOWES AND SONS, LIMITED, BECCLES.

Lightning Source UK Ltd.
Milton Keynes UK
UKHW022210201118
332627UK00015B/1620/P